Milk Stories: A History of the Dairy Industry in British Columbia 1827–2000

MILK STORIES

A History of the Dairy Industry in British Columbia 1827–2000

"HERE COMES THE TRUCK"

by K. Jane Watt

for the Dairy Industry Historical Society of British Columbia
Abbotsford, BC, 2000

Canadian Cataloguing in Publication Data

Watt, K. Jane
    Milk stories: a history of the dairy industry in British Columbia, 1827-2000.

Includes bibliographical references and index.
ISBN 0-9687663-0-7

    1.Dairying–British Columbia–History. I. Dairy Industry Historical Society
of British Columbia. II. Title

SF233.C2W38 2000   338.1'77'09711   C00-901087-4

Book Design: William Glasgow, Abbotsford, BC.

Printed in Canada by: Fraser Valley Custom Printers, Chilliwack, BC.

Cover Photo:
Cream Shippers, unknown location in the Fraser Valley, 1921.
Dairyworld Collection.

Title Page Art:
"Here Comes the Truck," by Diana Ponting, 1984.

Contents

President's Message

Dairy Industry Historical Society of British Columbia (left to right): Frank Bradley, member; Jane Watt, historian; Sandi Ulmi, member; Neil Gray, treasurer; Walter Goerzen, president; John Pendray, vice president; John Durham, secretary-manager; Gerry Adams, director. Missing members: Andy Dolberg; Earl Jenstad, Ken MacKenzie, and Allan Toop.

Milk Stories is the realization of a dream that began a number of years ago. As early as 1977, Everett Crowley, then president of the BC Dairy Council, accepted a motion that he write a history of the BC dairy industry. The material he wrote was first printed in *Butter-Fat* magazine in 1982-83 and included in *The Story of Avalon Dairy Ltd*, published by his wife, Jean M. Crowley, in 1996. In the early 1990s, several people again recognized the opportunity and the need to compile a history of the province's dairy industry. Stories were written and interviews recorded with the hope that somehow, sometime, a book would be published to include the material they had gathered.

This interest led to the incorporation of The Dairy Industry Historical Society of British Columbia on February 22, 1999. The Society formation was spearheaded by Frank Bradley, John Durham, Neil Gray, Jane Watt, and notably, Peter Friesen who became its first president.

The purpose of the society is to ensure that the fascinating history of the dairy industry be preserved. The primary focus to date has been the daunting task of publishing *Milk Stories*. In addition, the Society's mandate is to archive the wealth of material discovered and collected during the researching of the book.

Milk Stories is dedicated to all the men and women who were part of this great industry over the period of

**Peter Friesen
1925–1999**

history from 1827 to 2000. It is dedicated to all those who farmed the land, looked after the herd, milked the cows, transported the milk to the dairies, and worked in processing, bottling, marketing, and delivery to the consumer. The one individual who best typified all of these men and women was Peter Friesen. He was a dairy, poultry, and fruit producer, as well as an industry leader, serving as long term President of the Fraser Valley Milk Producers' Cooperative Association. In addition, he served on numerous other boards and committees in the agriculture industry and in his local community. After a courageous battle with cancer, Peter passed away on December 31, 1999 — an untimely death. His vision and leadership were key to the formation of the Dairy Industry Historical Society. Peter was a dairy industry statesman and he is greatly missed.

Milk Stories is about people, the communities in which they lived, and about the historical events that shaped the province and the industry. It includes merely a sample of the industry's rich pictorial record with photos from all regions of the province and all aspects of the industry. But, the most significant parts of the book are the interviews with the individuals who shaped this complex industry.

I wish to offer sincere appreciation to:

- The sponsors: companies, organizations, and individuals who believed in the project and helped make it possible by their financial support.
- Jane Watt, Ph.D., our historian, researcher, and editor, who did outstanding work in compiling the book.
- John Durham, who served as secretary, manager, administrator, fund-raiser, and who provided continual inspiration to the board.
- All members of the board, who gave direction and worked tirelessly to complete the project.
- Several editors, who proof-read and re-read the manuscript, notably, Sheila Durham, Neil Gray, Earl Jenstad, Tom Low, Barbara Souter and Sandi Ulmi.
- And, the many individuals who offered encouragement, advice, interviews, photographs, and information to make this publication what it is.

We recognize the outstanding contribution of the dairy industry to the development of the farming communities and the food industry of Canada. Today, less than three percent of Canadians are directly involved in agriculture, but our agrarian roots are the foundation of our future. *Milk Stories* is one way to acknowledge and celebrate the history and to educate many about the vital role of agriculture in the development of our country and our food supply.

It has been a distinct privilege for me to be part of this project, and I hope you enjoy *Milk Stories*.

Walter R.J. Goerzen, P Ag,
PRESIDENT
DAIRY INDUSTRY HISTORICAL
SOCIETY OF BRITISH COLUMBIA

Hector and Sarah Toop, pictured here with three of their seven daughters, one of their two sons, and three grandchildren. Chilliwack, 1905. It was through the dedication and hard work of many families like this one that dairying was established in British Columbia. Family members from left to right: Hector Toop, Mabel Toop, Albert Toop, Sarah Toop, Eugene Bennett, Cecily Bennett, Verna Pearson, Sarah Potts, Sophia McDermid.

Dedicated to the pioneers of the dairy industry

Preface

Many stories told about British Columbia concentrate on its rich resource heritage and on the people who were a part of that heritage, such as miners who worked the Fraser and followed Billy Barker east of Quesnel to the new town called Barkerville, or took a chance on getting rich beneath the mountains of the Kootenays or the eastern slopes of Vancouver Island. Other stories are of the bush, of fallers, buckers, chokermen, and rigging-slingers — loggers working to bring the wealth of wood to market. Many are about isolation and living in the moment, about long shifts and temporary living in mining or logging camps, about the relief of getting to town to party or home for a spell to see the wife and kids, about returning to camp to do it all again. These resource-based stories, even myths, foster images of rugged independence, reinforcing the idea of the province as a wild west.

The story of the dairy industry in British Columbia is a different kind of story. It is not about isolation; it is about community. It is not about exploiting resources; it is about fostering growth on the land. It is about stewardship and the importance of daily routines. It is about planning and working hard at jobs in the present to guarantee the future. It is about settlement, about working to make vision a reality. Staying put, milking twice a day, keeping meticulous records, and valuing the contributions of all members of the family might not seem as exciting as logging camp adventures or prospector's dreams, but they are as important. They demonstrate a commitment and an optimism that have been essential to the history of British Columbia. But calling them stories of settlement doesn't mean that they are dull. Some are funny. A few are sad. They are stories about luck, hard work, stubbornness, determination, cooperation, and sometimes desperation.

The story of the dairy industry is important to British Columbia not only because of the quantity of milk the industry produces, but also because its network touches the daily lives of British Columbians in fundamental ways. Each day, all year, 1.6 million litres of milk moves from BC farms to processing plants. Milk and milk products move from processing plants to urban doorsteps and to thousands of stores, making the production and movement of dairy products a vital part of the rhythm of daily life in the province. But its importance to the province can't be measured only in litres or in dollars. People depend on milk. Years ago, city dwellers knew that they had only to open their front door in the morning to find fresh milk. They knew, too, that in bad weather, or in times of shortage, the milkman would look out for the community, making sure families with children or invalids were served first and all others were served as soon as possible. Today, people know milk will be fresh, stocked on the shelves of their local supermarket or corner store, wherever they live in the province.

This is a book about people — their ideas and inventions, their changing needs and expectations, and their commitment to service. While the dairy industry in British Columbia today relies on approximately 760 producers, it was for many years supported by many thousand more small farmers and farming families living mostly in six geographical areas: the Fraser Valley, the Okanagan, the Kootenays, the Peace River area, the Bulkley and Nechako valleys, and Vancouver Island. The history of dairy farming is tied to the history of these areas. And the reverse is also true: the history of these areas is tied to the history of dairy farming.

This book is a only a beginning. Our hope is that through this beginning, dairy producers will be inspired to compile the histories of their farms and their localities and that consumers will be challenged to reflect on the importance of agriculture to their daily lives, and on the monumental task — both in the past and today — of feeding a population. ∎

First Farms, First Technologies

Officials from BC Electric, the Langley Farmers' Institute and the Department of Agriculture demonstrate a new method of burning stumps. The method employed large electric fans, waste oil, and a good deal of patience. Belmont Farm, Langley, early 1930s.

BUTTER-FAT, 1932

The first recorded dairy farm in what is now the province of British Columbia was the Hudson's Bay farm near the present town of Fort Langley. In the early 1800s the North-West Company had set up simple trading posts with large gardens and a few livestock at Fort St. James, Fort McLeod, Fort Fraser, Fort George, and Fort St. John to mark trading territory and to muster furs. But it was not until after the amalgamation of the North-West Company and the Hudson's Bay Company in 1821 that the establishment of farming operations became a priority. In 1824, an exploration party of 40 men under Chief Trader James Macmillan travelled by canoe from Boundary Bay up the Nicomekl River, portaged to the Salmon River, and followed it to the Fraser River. They travelled upriver as far as Hatzic Slough before turning and following the river to its mouth. They were searching for a place to establish the presence of the Hudson's Bay Company (HBC) to discourage American and Russian traders who plied these waters. They were also looking for a tract of land suitable for farming to fulfill a new company policy that called for less extravagant living and an end to the company's reliance on imported goods; in short, under the direction of HBC Governor George Simpson, "living was to become much plainer and the posts made more self sufficient."[1]

They must have been satisfied with what they had seen on the southern shores of the Fraser because in 1827 Fort Langley was established. George Barnston's journals of 1827 and 1828 tally the labour involved in clearing the land. He writes that the work is "of a very labourious description, the timber being strong and the

Advertisement for the sale of livestock from the Hudson's Bay Company farm. *Mainland Guardian,* **April 5, 1871.**
VANCOUVER PUBLIC LIBRARY

Advertisement for the sale of the Hudson's Bay Company farm, *Mainland Guardian,* **December 19, 1877.**
VANCOUVER PUBLIC LIBRARY

New Westminster, 1906.

Devonshire Cream

Mother was great for making Devonshire cream. That is made by placing a pan of milk — when the cream is all up to the top after it has been setting for say 24 or 36 hours — on the back of the stove, until just before it comes to a boil. And when you'd see that the cream was all wrinkled and sort of moving, twitching like there were nerves in it, she would know it was ready. Then she'd set it in a cool place until it got perfectly cold and this cream would be solid, and you could just scoop it off. And it was the most delightful thing I ever tasted.

Fred Toop, 1963.

ground completely covered with thick underwood, which is closely interwoven with Brambles & Briars."[2] In 1828, the ground was broken for potatoes. The following year, according to Archibald McDonald, "of about 15 acres now open, five of them are low meadow — five fine mellow ground fit for the plough, & the rest full of Strong Stumps & root fit only for the Hoe for many years to come: and pasture for more than a few Beasts, is out of the question."[3] But the "beasts" arrived the following year.

Eventually, the main farming operation was moved from the banks of the Fraser River to two thousand acres of land on Langley Prairie, on the lands between the Salmon and the Nicomekl rivers. In 1839, the Hudson's Bay Company entered into an agreement with the Russian American Company (RAC) to lease a large portion of the Pacific coast for 10 years, thereby securing a trading monopoly there. The Hudson's Bay Company agreed to provide the Russian American Company with two thousand seasoned land otter per year as well as "to transport from England British manufactured goods desired by the Russian colonies, to sell the Russians additional land otter and to provide supplies of foodstuffs, including wheat, peas, barley, butter, beef and ham."[4] With the signing of this lease, agricultural products, especially butter, became important commodities. In 1839, a new Fort Langley was built two-and-a-half miles upstream from its original site; a place, according to James Douglas, "alike convenient for the fur and salmon trade, combined with facilities for the farm and shipping."[5] This interest in the fishery and farm marked a change in HBC strategic planning: the

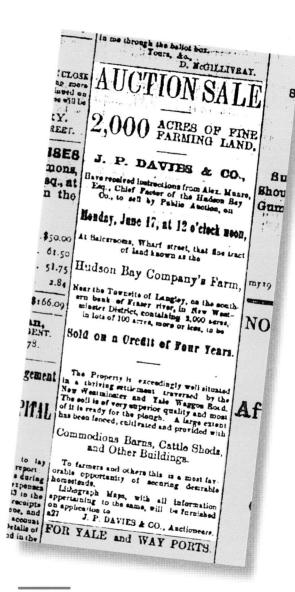

Advertisement for the sale of the Hudson's Bay Company farm, *Mainland Guardian,* **June 15, 1871.**

Panorama of Higgins farm in Eburne, taken from the foot of Granville, 1908.

production of foodstuffs — wheat, butter, and salmon — were no longer considered peripheral to the fur trade, but rather as valid company ventures in their own right. In 1839, Chief Factor Dr. McLoughlan brought 29 milk cows to the Fort on board the *Beaver* "to augment the Fort Langley herd"[6] as well as an "English family to take charge of one of two dairies to be established immediately."[7] When fire destroyed the second Fort Langley site in 1840, it dealt a serious blow to the company. It relied on the production of Fort Langley not only to ship food supplies to other forts, but also to produce surplus to service its lease.

The importance of the dairy industry to the newly diversified economy of the company is illustrated in a story told by Chief Trader Ovid Allard many years later. Despite Allard's reputation for exaggeration in storytelling,[8] it is instructive. Apparently, a woman named Mrs. Findlay was in charge of the dairy at the time of the fire in 1840. Being very proud of her butter, Mrs. Findlay ran around as the fire raged, "appealing to the men to save the cream, and in despair, ran into the burning buildings herself and brought it out. Not until that important matter had been attended to did it occur to her that her infant child was still within the blazing structure, and, while her offspring was being rescued, the fire reached the cream and ruined it. 'Oh! dear,' cried Mrs. Findlay, 'what a fine batch of butter will never be made now.'"[9] The Hudson's Bay post at Fort Langley was important to the company, but "was not the usual type of Post founded by the company, inasmuch as the fur trade in that territory was negligible,"[10] according to James Robert Anderson, the first perma-

nent official in the provincial Department of Agriculture. "But it attained importance as a point where provisions could be obtained for the remainder of the Posts along the Pacific Coast. The curing of salmon and oolichans, the growing of potatoes and fresh vegetables and the supplying of dairy products, were the business of Fort Langley at that time. Quite a large dairy herd was maintained at the Post for the latter purpose, the words 'Picked from the Langley herd of cattle' appearing quite frequently in whatever advertisements there were of stock for sale."[11]

Shifting political and economic forces required the Hudson's Bay Company to move its headquarters on the Pacific coast to Fort Victoria on Vancouver Island from Fort Vancouver in Washington. Construction began in 1843 "using Indian labour paid at the rate of one Hudson's Bay blanket for every forty cedar pickets cut."[12] Three farms were established near Fort Victoria: the Fort Farm on the flat where the business section of Victoria now stands, Beckley Farm, south of James Bay, and the North Dairy Farm.[13] Cows "of Spanish origin, obtained from the Mission Fathers in California"[14] had been driven overland to the Columbia River and dispersed to Hudson's Bay forts at Nisqually and Fort Vancouver. After the Treaty of Washington was signed in 1846 and the international border was established at the 49th parallel, the Hudson's Bay Company retained trading and trapping rights south of the new border, but it concentrated on establishing a strong presence and a lucrative operation on what was now Canadian soil. Provisioning its northern posts through its posts in the Columbia basin was no longer an option, as the

Cow crossing the road in Eburne, now the approximate location of Vancouver's Oak Street Bridge, 1908.

The Importance of Butter

She took on the chore of milking the nine cows night and morning because her husband, besides clearing and improving his land, had undertaken the job of superintendent of roads for the Municipality of North Cowichan. Besides caring for her small children, cooking and sewing, Margaret made the butter and looked after the poultry and the vegetable garden. The butter and egg money went a long way in providing groceries. She made butter for sale and it would be packed in specially made boxes, taken to Maple Bay and shipped to Victoria. She not only churned her own butter, she also prepared for sale butter churned by her bachelor neighbour ... Until 1886 they had to spend the better part of a day once a week taking the team and wagon to Maple Bay to meet the boat, to ship butter and eggs ... Once the railroad came through in 1886, life was much easier as Somenos station was only a mile away. Then the butter and eggs could be shipped by express to Nanaimo. As soon as the children were old enough, they had to help with the chores, such as milking. There were no fences yet and the cows wandered far and wide in the woods. The three eldest daughters were able to help their mother with cooking, housework, and the care of the younger children and the boys were able to take over the milking. But Margaret still supervised the baking, and made the butter.

Memories Never Last: Stories of the Pioneer Women of the Cowichan Valley, 1986.

A Woman and a Cow Reach the Cowichan Valley

In the year 1865 some disembodied spirit might have observed a middle-aged woman trudging the rough Indian trail from Victoria to Cowichan. Her progress was slow as she was allowing the cow in her care to graze along the way. The rest of the party had gone ahead and made an early camp, and she was to catch up with them at the end of the day. This Elizabeth Blackmore was said by some accounts to have been the first white woman to *foot it* to the Cowichan Valley. While the route taken was roundabout, via the Sooke Lakes and Shawnigan, it was the only possible one at the time.

Memories Never Last: Stories of the Pioneer Women of the Cowichan Valley, 1986.

company was not only subject to duties on goods moved north but also on furs and trade goods moved south. Following the completion of new company headquarters in Victoria, cows were brought from forts in Washington and Roderick Finlayson was appointed manager. In 1846, 140 cattle were supplying the needs of the inhabitants of Fort Victoria.[15] By 1860, five milk sellers were included under "trades and professions" in a city directory, along with a whole range of others, including "seven auctioneers and estate agents, seven dealers in tin and hardware, three wholesale liquour dealers, six cigar and fruit dealers, two ship builders, one patent roofing dealer, four lodging houses, one tanner, one underwriter, one crockery and glass dealer, one furdresser, two seed dealers, two breweries ... two saddlers, and eight wharves."[16]

The discovery of gold in the Fraser River in the spring of 1858 had important effects on the development of agriculture in British Columbia. Suddenly, the populations on the western edge of the continent — in the British colony on Vancouver Island and in the separate colony of New Caledonia (renamed British Columbia later that year) — were on the move east and north into the gold fields in the Fraser and Thompson rivers. And they were not alone. Along the coast of the United States, preparations were being made, especially by the "49ers," the veterans of California's gold fever. In San Francisco, the newspaper lamented the exodus of workers to the far off fields of New Caledonia: "There is no telling when this system of depopulation is to stop. At the present time, the boats from the interior come down every night, loaded with miners and others, all bound

THE NEW ECONOMY CHIEF CREAM SEPARATOR MARVEL of 1908 $28.80

OUR LATEST AND BEST. THE ENVY OF ALL OTHER SEPARATOR MANUFACTURERS. MORE SIMPLE THAN OTHERS, MORE EFFICIENT THAN OTHERS. THE REAL WONDER OF THE CREAM SEPARATOR WORLD.

MONEY CANNOT BUY A SEPARATOR that will compare with the new Economy Chief for 1908. It is built in by far the biggest separator factory in the world by our own mechanical experts, picked men, the cream of the country, whose reputation is established wherever high grade machinery is made. It far surpasses any other machine offered by any other maker. No other maker has the facilities we have. No other maker has the same amount of money invested in specially designed machinery built for the sole purpose of constructing perfect cream separators.

FIRST OF ALL the new Economy Chief is a skimming wonder of wonders. It skims to the last small drop of cream under the hardest conditions. It skims every particle of cream where all others fall, clog up and become useless.

THE NEW ECONOMY CHIEF FOR 1908, we now offer to for the first time at only $28.80 to $43.65, prices below the actual cost to produce separators in other factories and only possible by combining the facilities of the biggest and most complete cream separator factory on the globe with our great and unapproachable selling power, the power to reach and put out goods before the eyes of practically every farmer in the United States at a cost so small it is only nominal on each sale.

WE FULLY GUARANTEE TO YOU to say nothing of our latest and best, the new Economy Separator, when run by yourself or anyone else competent to run a cream separator, will outskim any $125.00 separator on the market two to one on your own farm.

IT HAS A DOZEN NEW AND VALUABLE IMPROVEMENTS.

THE CRANK IS JUST THE RIGHT HEIGHT, in the position recommended by medical authorities as the most natural and healthy.

IT RUNS SURPRISINGLY EASY. The new Economy Chief is the simplest possible separator.

MANY OTHER IMPROVEMENTS aid in making the new Economy Chief the most perfect separator ever produced.

THE MOST SIMPLE SEPARATOR ever devised.

IT WILL LAST LONGER, from three to ten times as long as other separators.

OUR PRICES AS GIVEN BELOW ARE FACTORY COST with our small profit added and nothing more.

FACTORY TO YOU PRICES FOR THE NEW 1908 ECONOMY CHIEF CREAM SEPARATOR

No. 23K61 The New Economy Chief Cream Separator for 1908, with all the latest improvements, capacity 250 to 300 pounds per hour. Shipping weight, 185 pounds. Special price to introduce on sixty days' trial........ $28.80

No. 23K62 The New Economy Chief Cream Separator for 1908, with all the latest improvements, capacity 350 to 400 pounds per hour. Shipping weight, 195 pounds. Special price to introduce on sixty days' trial........ 34.95

No. 23K63 The New Economy Chief Cream Separator for 1908, with all the latest improvements, capacity 600 pounds per hour. Shipping weight, 205 pounds. Special price to introduce on sixty days' trial........ 43.65

For Quick Delivery and Low Freight Offer on Our Large Economy Chief Separator, see page 34.

Advertisement for cream separator.   SEARS ROEBUCK AND COMPANY CATALOGUE, 1908.

WHAT DO YOU KEEP COWS FOR?

On the road to the creamery. A single can carries all the cream from a wagon load of milk if you separate it at home with the Economy.

NEARLY ALL FARMERS KEEP COWS TO MAKE MONEY

DAIRYING IS ONE OF THE MOST IMPORTANT FARM INDUSTRIES AND IS BECOMING MORE IMPORTANT EVERY DAY. A LARGE PART OF THE AVERAGE FARMER'S ENERGY AND TIME IS GIVEN TO THE CARE AND FEEDING OF HIS COWS. THE MATTER ANY THOUGHT HE KNOWS THAT HE DOES IT IN ORDER THAT HE MAY MAKE MONEY FROM THE PRODUCT OF THE COWS.

IF YOU KEEP COWS a considerable portion of your land is no doubt either to raise forage crops or for pasture. You get up early in the morning to milk and feed them and in the evening you milk them again.

YOU KEEP COWS PRINCIPALLY FOR THEIR MILK. The most valuable part of the milk and, in fact, the most valuable of all farm products is the butter fat.

BUTTER FAT IS SO VERY VALUABLE, so much higher in price than any other farm product that if any considerable part of it is wasted the farmer not only makes no profit but actually is a loser.

IF YOUR COWS ARE GOOD and intelligently selected for their milking qualities, properly fed and cared for and if you get all the butter fat, the two-thirds is your profit, you will surely prosper.

THE EXTRA COST OF A FINE DAIRY COW, even if it be two or three times as much more than pays her board in butter fat.

THERE ARE A NUMBER OF WAYS of taking all or part of the butter fat from the whole milk for general use.

THE BEST WAY OF ALL and the one which is now being very generally adopted is to separate the cream containing the butter fat from the whole milk by means of the hand separator.

RAISING CREAM IN PANS AND CROCKS DOESN'T PAY

IF YOU HAVE BEEN RAISING CREAM IN CROCKS AND PANS in the old fashioned way we feel safe in saying that your dairy does not pay.

THERE IS VERY LITTLE DIFFERENCE in specific gravity between butter fat and skim milk.

THE CREAM THAT HAS RISEN to the surface has stood for hours exposed to the action of germs, and has perhaps absorbed musty odors which are always present in the cellar or spring house.

THE LOSS IN THE PAN RAISING SYSTEM is not always one-third.

THE HARD WORK AND SLAVERY of the pan raising way of gathering cream.

THE RESULT OF IT ALL is that one-third of your butter fat, which is all your profit, goes to waste in the sour skim milk.

THE DILUTION SYSTEM, by which a large quantity of cold water is mixed with milk.

THE DEEP SETTING SYSTEM of raising the cream in deep round or oval cans plunged in ice water.

A DISADVANTAGE OF ALL GRAVITY RAISED CREAM is that it will not churn completely.

THIS IS THE HARD, OLD FASHIONED, MONEY LOSING WAY of dairying.

Advertisement for cream separator.   SEARS ROEBUCK AND COMPANY CATALOGUE, 1908.

Cheese

James McIntyre (1827-1906) lived in Ontario and was known as "the cheese poet" because he devoted so much poetry to celebrating the production of cheese and the importance of the dairy industry in Canada. Surprisingly, he was not a farmer, but was a furniture maker, an undertaker, and a teacher of elocution. He was frequently called upon to put together a few words for a range of occasions. Here are some selections from his work.

From "Hints to Cheese Makers"

All those who quality do prize
Must study colour, taste and
*   size,*
And keep their dishes clean and
*   sweet,*
And all things round their facto-
*   ries neat,*
For dairymen insist that these
Are all important points in
*   cheese.*

From "Lines Read at a Dairymen's Supper, 1899"

Then let the farmers justly prize
The cows for land they fertilize,
And let us all with songs and
*   glees*
Invoke success into the cheese.

Gibson Brothers' Gold Bank Dairy Farm, Jura, 1910. Jura was a flag station on the Kettle Valley Railway line, 10 miles east of Princeton.
BCARS B02760

for Fraser's River. The hotels of this city are fairly crammed with people waiting for an opportunity to leave … Throughout the length and breadth of the State, the 'Fraser River Fever' seems to have seized hold of the people, and threatens to break up, or at least seriously disarrange for the time being the entire mining business of the State."[17]

This influx of miners was an overwhelming change to the sedate colony. For example, "by 1854 there were only 250 people living in the farms around Victoria … Then, on Sunday, April 25, 1858, the people of Victoria came out of church to find the paddle steamer *Commodore* had just docked from San Francisco with 450 min-

ers eager to start looking for gold."[18] It is estimated that almost 30,000 people reached the Fraser that year.[19] Not only did settlements grow overnight, but they also fundamentally changed character: "British subjects suddenly found themselves jostling native-born Americans, blacks, Chinese, Germans, Italians, Jews, and Spaniards on the streets of Victoria."[20] En route to the gold fields that gradually extended north until they reached the Cariboo and the large finds at Barkerville, these miners needed not only maps and directions, but also food. And the prices of food were dependent on distance: a bag of flour, for example, "worth $16 in Whatcom, sold for $25 at Fort Langley, $36 at Fort Hope, and $100 at

Dairy Farm near Victoria, circa 1900.

Making Butter, Using a Cream Separator

To make butter in those days we used to set the milk in pans. Some people had deep creamers that cooled it, then they'd drain the milk off at the bottom till they got down to where the cream would show in a glass. But we had it set in pans in a very cool dairy that was built into the log cabin. Mother would skim the cream off, and when it was ready to churn, she put it in this wooden dash churn. I've sat in the corner, before I was ready to go to school, and churned butter many a day with one of those wooden dashes ... Then Mother would make that up at a nice butter table where she used a lovely roller, an octagon roller that I think my oldest brother must have made for her.

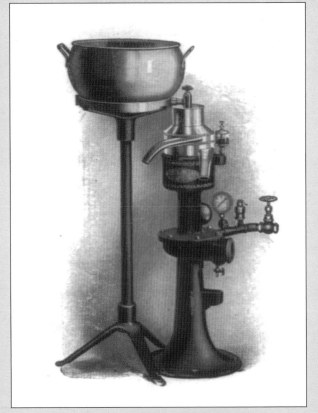

Steam powered cream separator. From a De Laval Dairy Supply catalogue, 1912.
ACTON KILBY PAPERS, KILBY FARM MUSEUM

She'd make this mould right up with butter; and it had a plunger at the top, and you'd push down on that, and it would push that pound of butter out onto a table. And she would wrap it up in a piece of special paper — waxed, I guess it would be — that would make it air-proof; and of course it would keep in that for quite a long time. Then she would take it into Chilliwack and sell it to the stores for products. That was one of the ways we had of getting a livelihood. She could always sell all the butter she could make. Then the skim milk, of course, was fed to the calves and pigs.

We went from that to a little separator called an Alexandra cream separator. One could milk about as fast as that thing would separate it. The question was what to do with the cream. We were getting more cows all the time. So we shipped the cream by boat to New Westminster to a creamery in a big barrel that had a tin lining in it. We would ship it there when it had pretty well filled up, because it was made into butter and didn't need to be sweet, as long as it was stirred up and kept in a cool place so that it didn't go bad in any way. These cans had a heavy lid on them and a padlock put on it; and we had a key and they had a key down there at the dairy. It would be taken to the riverboat and to New Westminster in one day, and the next day it would come back empty. And that's the way we did it for quite a while.

Fred Toop, 1963.

the farthest gold diggings. While it cost just over $20 to dispatch a ton of freight from London to Victoria, the charge from Victoria to the farthest reach of the gold fields surpassed $200 by 1859."[21] It is estimated that between 3,000 and 5,000 miners spent the winter of 1858-59 in the Fraser Valley, most between Fort Langley and Fort Yale.[22]

"At this time," according to Edward Philip Johnson, "land was relatively easy to obtain." Johnson writes about the early history of Ashcroft Manor, and the ranching experiences of the Cornwall brothers in the Cariboo: "A settler could purchase a quarter section for about the same price as a good horse. By the Proclamations of January 4, 1860 and January 19, 1861, unsurveyed land claims in British Columbia, 160 acres in size and rectangular in shape, could be pre-empted by British subjects or by anyone else who took an oath of allegiance to the Queen. The settler had to live on and improve the land, register his pre-emption claim with the district magistrate, and pay a two dollar registration fee. He could obtain clear title to the land by getting a certificate of improvements from the magistrate and paying one dollar an acre for the land. In addition the settler could buy unoccupied land adjoining to his 160 acre claim for the same price per acre."[23]

Farms began to take root not only on the banks of the Fraser, but also on Saltspring Island, in the Cowichan Valley, in the Okanagan, in the Thompson and Nicola valleys, and in the Kootenays. Large ranches were established in the interior of the province, including the Cornwall in 1862, the Australian in 1862, the Bonaparte in 1865, the Harper Ranch in 1869, the

Farm at the head of Cowichan Bay, circa 1900.

Toop family herd, Chilliwack, circa 1900.

ALLAN TOOP COLLECTION

Vernon (later the Coldstream), the BX, and the O'Keefe. These large establishments were like small towns in their own right and developed a high degree of self-sufficiency — employees worked not just as cowboys in the drive to produce beef, but also in the dairy, in the garden, and in the kitchen to provide for the large crews that kept the ranches going.

In 1865, the Chadsey Brothers of Chilliwack produced 3,000 pounds of butter.[24] In an often recalled transaction in 1868, the Chadseys recorded selling 2,500 pounds of butter to the "Cariboo Gold Diggings." Six thousand pounds of butter from the Sumas district generally was sold by one dealer,[25] retailing in Barkerville for one dollar per pound.[26] James and William Chadsey pioneered not only different methods of

farming in the Chilliwack area, but also new marketing methods for their products. They brought the first grain mill to the valley, built a grist mill in Sumas, brought two bulls from Oregon, and in doing so introduced the first Holstein-Friesians to the province. Their butter was shipped to the gold fields in two, five, and ten-pound air tight cans.

Early reports listed Vancouver Island as a prime farming area. Indeed, Hibben's *British Columbia Directory* of 1877-78 notes of the Cowichan area that "under a judicious system of farming there can be no doubt but that good returns can be obtained from these lands as from any part of the continent of America."[27] In Duncan, for example, "prior to 1900, when the town's population figures hovered in the high double digits, the engine of

Heffley Creek Creamery, 1916.

Cream Separator

In the early days we used the dash churn. It was some forty-six years ago that I got the first barrel churn in the district. Few settlers had any dairy house or building just used to store milk — the milk was kept anywhere. It might be placed alongside a barrel of fish or other food. The skimming of the milk was a slow process which took the pioneer women a long time to accomplish. Our first advance was to the "deep setting can" which held three gallons and was set in cold water at 45°F for twelve hours. This was a saving of time and labour.

The settlers used the old upright churns for butter-making, and in early days printed it in round prints much the shape of a saucer, without any mould to shape it. The same prints came later with a mould, but were not adapted to shipment for any distance, and we wrapped the prints in cheesecloth. That was before we got wax butter paper. Then came the round mould of two pounds and after that we shipped in blocks ...

Our steamboat service was in many ways ahead of the railway. A ship would take our full box of butter and charge for it, but bring back the empty box free of all charges. Some settlers who only milked a few cows sent their butter to the local stores and took it out in trade. The two stores in Victoria which took the greatest amount of Cowichan butter were Henry Saunders on Johnson Street, and James Fell on Fort Street. The Cowichan Creamery was formed and installed a cream separator. The creamery had milk wagons hauling the milk to the creamery separator, and then hauling the skim milk back to the farms. The haulers only got about $35 per month for team and wagon. In a short time this was changed and each farmer installed a cream separator at home and took the cream to the creamery.

The cream separator put dairying where it stands today. Because of it the many creameries developed, and the Cowichan Creamery has been the greatest blessing to come to the Cowichan district.

John N. Evans, 1936.

prosperity was agriculture. All around Duncan, men were clearing the land and planting crops. There seemed to be no end to what the valley's rich soils and warm wet climate would produce: tobacco, grapes, potatoes and especially grass. Thick, sweet grass grew nearly year-round in the valley, reducing the need to lay in huge crops for winter feed. The local climate was said to add $70 a hectare to the value of farmland. There were more dairy cattle in the Cowichan Valley than anywhere in BC except the Fraser Valley. Milk, cream and butter from cows fed on this rich pastureland were sought all over the island."[28]

Sometimes these returns in the form of vital and prosperous settlements were strongly linked to the work of individuals. In the Comox area, for example, the development of dairying was tied to the work of Alex Urquhart, who arrived in the Comox Valley in 1871. Other farmers were already there, like Duncan, McNish, Wilson, Robb, Fitzgerald, and Thompson. McNish was milking Shorthorns and he sold some of his best cows to Urquhart in the summer of 1871. By 1877, this area was described as "an agricultural settlement, prosperous and contented."[29] By 1885, Jerseys had come to the Comox Valley; indeed, Urquhart was milking about 40 cows at this time and marketing his butter to the Al Johnson company of Nanaimo.

In this area, butter was first put up in two pound sacks sewn by local women. Later, the two pound roll was made popular by imported California butter, and the sack was replaced by this popular roll. Butter from the Comox Valley was sold to miners in neighbouring areas of Nanaimo, Wellington, and later Cumberland,

A.C. Wells and son, Edenbank Farm, Chilliwack, 1895.

Early Days at Edenbank

Around Allen and Sarah's cabin (the home of A.C. Wells and his wife, Sarah) a vegetable garden grew in the clearing. Potatoes, beans, and carrots, wallflowers and tall hollyhocks stood against the house. At night a coal-oil lamp burned in the window so that Allen might more clearly see his way up from the barn ... At daybreak the work began. Allen toiled alone ... The brush was slowly cleared from the meadowlands where pasture, marshy though it was, was sufficient for the cattle in the field. When the cows calved and there was milk in quantity available, great skimming pans were set in a cool and shady spot while the cream rose to the surface. Sarah would skim it off and churn it into butter. The skim milk went back to the calves, or to the pigs in the pen behind the barn. At night, the hogs were closed within their sty to protect them from bears and wolves which lurked in the surrounding woods at night. The cattle were a heavy breed, mostly of Durham blood, originating on the Hudson's Bay Company farm at Fort Vancouver by the Columbia River. Females were raised for milk production and the males were castrated to become beasts of burden on the Cariboo road, or used for meat in the beef-hungry gold fields of the interior.

Oliver Wells, 1967.

A.C. Wells and Sarah Wells, Chilliwack, 1880s.

principally through the Hudson's Bay Company store. The steamer *Maud* and the steamer *Cariboo Fly* "also afforded other outlets for produce of all descriptions" and J. Wilson, the trader on these boats "was well known on the East coast of Vancouver Island by all of the old timers."[30]

In 1891, the Dominion Census listed 6,500 farmers living in the province.[31] Most of these were mixed farmers — they didn't specialize, but instead produced a range of different products, including milk and butter, to feed their families and then marketed the surplus at local stores. Others, however, specialized early on and the development of the centrifugal cream separator in 1878 suddenly made it quicker and easier for these farmers to process a greater volume of milk than they could do by hand. By 1885, A.C. Wells of Chilliwack had immersed himself in learning this new technology through books, had travelled to Ontario to hire an expert cheese and butter maker, and had opened a creamery to process his own Edenbank Farm milk. It didn't take long for the Chilliwack area to establish a name for itself: the *British Columbia Directory* of 1892 proclaims Chilliwack to be a place noted for its excellent butter, reputed to be full of flavour, and made in large quantities. Under Wells' direction, the five-member Edenbank Creamery was established in 1896, the first farmer's co-operative in British Columbia. By 1901, it boasted 72 members.

Another important early technology was the Babcock Tester, a device invented to measure the butterfat content of milk. It allowed farmers to begin to keep detailed production records. Developed, not surprisingly, by Dr. S.M. Babcock, the test was "based on the fact that strong sulfuric acid will dissolve all non-fatty solid constituents of milk and other dairy products, and thus enable the fat to separate on standing."[32] In order to test a sample of milk, the tester placed the bottle holding the mixture of milk and acid into a centrifugal machine and whirled the sample for four minutes. Hot water was

A portion of the milking herd of Mr. Thompson, Terra Nova (now part of Richmond), circa 1890.

R.U. Hurford

The First Cream Separator in the Comox District, 1887

The first cream separator was purchased from our old friend, the late Mr. Wells of Chilliwack, about 1887 by Mr. Urquhart, although he was very sceptical about a machine's being able to separate cream from milk. He carefully weighed the milk for two days previous to trying out the separator and for two days after, and found that the actual churning of the cream yielded 11 pounds more butter. When the news circulated around, a steady influx of visitors came to see this wonderful machine, and soon an agent for cream separators was appointed in the district and they ceased to be a novelty.

R.U. Hurford, 1925.

added to bring the liquid fat into the graduated neck of the test bottles, the sample was whirled again, and the length of the column of fat, indicating the percentage of fat, was read. This test remained a standard test well into the second half of the twentieth century and, according to a 1911 manual called *Testing Milk and Its Products: a Manual for Dairy Students, Creamery and Cheese Factory Operators, Food Chemists, and Diary Farmers,* "any kind of milk can be tested by the Babcock Test. Breed, periods of lactation, quality or age of the milk are of no importance in using this method, so long as a fair sample of milk can be secured." [33]

In this early period, transportation of milk over great distances was difficult, if not impossible. Roads were poor or non-existent and the age of the railway was just beginning. Local areas developed markets and local sources of milk supply. Creameries also developed by the end of the nineteenth century, such as the Delta Creamery in Ladner, built in 1895 to serve the expanding metropolis of Vancouver. According to the *Commonwealth* newspaper from New Westminster, this proximity to Vancouver and good "land and climate" made dairying a "paying industry." In Richmond, for example, "the farms … have an air of comfort and settlement which is often wanting in those situated in the wooded sections. The large barns speak eloquently of heavy crops and the ample stable accommodation tells, without seeing the cows at all, that dairying is a paying business." [34] Prosperity for farmers depended, not just on quality products and fine herds, but on proximity to population. It would take the development of reliable roads and railways to change this equation. ■

Dr. S.M. Babcock, Inventor of the Babcock Milk Test. From *Testing Milk and Its Products: A Manual for Dairy Students, Creamery and Cheese Factory Operators, Food Chemists, and Dairy Farmers,* 1911.

Ad for the Babcock Tester. From *Old Time Agriculture in the Ads,* 1970.

Creamery at Port Hammond, 1907.

CHAPTER II

Early Government Support

A gathering in Victoria, likely a meeting of the Farmers Advisory Board of the B.C. Dairymen's Association, circa 1920s. Left to right: John W. Berry, Belmont Farm, Langley; Pete Moore, Colony Farm, Essondale; Henry Rive, Dairy Commissioner, Victoria; Edwin A. Wells, Edenbank Farm, Sardis.
BC FARM MACHINERY MUSEUM

The relationship between the support of government and the development of the dairy industry in British Columbia is a clear one. Through their combined workings, the federal Department of Agriculture and the provincial Department of Agriculture did much to assist early farmers. Through research, extension work, and partnerships with the new Department of Agriculture at the University of British Columbia, these agencies developed close, constructive relationships with farms and farmers. Both levels of government were active in anticipating the needs of early farmers in British Columbia and expanded offices and staff to accommodate the movement of settlers to remote areas of the province or to places where new transportation systems made the large scale movement of agricultural produce possible for the first time. Sometimes incentives to settlement were financial, such as the deferred payments and low interest loans offered to returning veterans of the First World War with farming aspirations.

The first Dominion Dairy Commissioner was appointed in 1890 and it was the mandate of his department to initiate programs and policies to support the development of the dairy industry across Canada. "Early federal initiatives included an iced butter railway car service (1895), funding for cool cheese curing rooms (1902) and the grading of butter and cheese for export (1923)."[1] Through its work at Agassiz Experimental Farm, the federal Department of Agriculture did much to assist early farmers who were "urgently in need of information that would help them grow crops on previously uncultivated soils and under climactic conditions

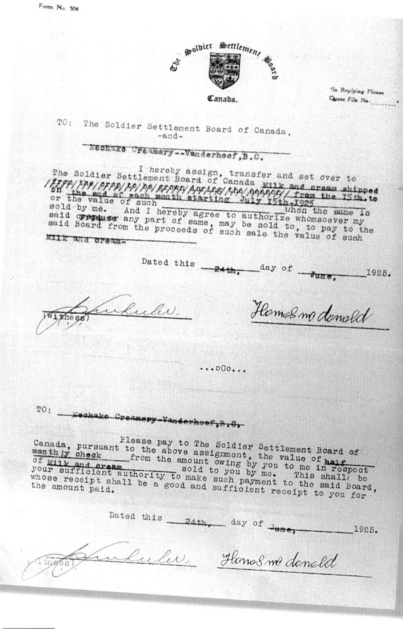

Contract between Soldier Settlement Board and Thomas McDonald outlining the arrangement that half of his monthly milk cheque from the Nechako Creamery in Vanderhoof be automatically forwarded to the Board, 1926. BCARS ADD. MSS. 1150

Jersey cow and handler, Saanichton Experimental Station, circa 1930.

Federal Support

Farming is a complicated business. Many of the problems connected with it are as old as agriculture itself and new problems are arising constantly. Today, more than ever before, progress and success in farming depend very largely upon the facts as discovered by science, and the application of these facts to the problems confronting the farmers. Every member of the staff has the interest of agriculture at heart and welcomes the opportunity to serve. The greatest good can only be accomplished if farmers of the country make the fullest possible use of the information and services which are available to them.

Preface to Fifty years of Progress on Dominion Experimental Farms: 1886-1936.

Dominion Experimental Farm, Agassiz, 1898.

BCARS B06838

only incompletely understood."[2]

In 1894, the Provincial Department of Agriculture was formed with the mandate to foster the growth of agriculture in the province. Not only was it a clearinghouse of information on land improvement and animal husbandry for early farmers, but it also passed legislation to guide the rights of producers and consumers.

Other support came from partnerships between government and communities: early farming associations supported dairy families by sponsoring speakers, field days, and exhibitions that contributed to the education of the farm community and allowed opportunities for the community to socialize. Associations, including the BC Dairymen's Association and regional branches of the Farmers' Institute and the Women's Institute, provided

forums for formal comparison of dairy animals and products while bringing the community together to talk and to do a little business.

Agassiz Experimental Farm

The Experimental Farm at Agassiz was one of five experimental farms provided for by the Government of Canada in 1886 when it passed *An Act Respecting Experimental Farm Stations*. These stations were "established to serve the farming community and assist the Canadian agricultural industry during its early development"[3] through research programs focussing on soil management, crop production, animal productivity, and food processing. By the fall of 1889, land was purchased in Agassiz and Mr. Thomas A. Sharpe was appointed

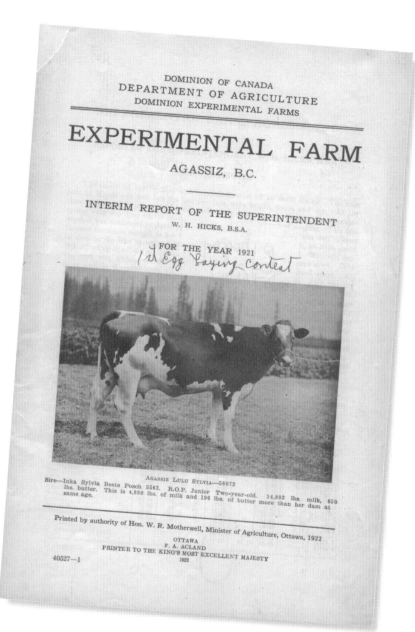

Cover, Interim Report for 1921, Dominion Experimental Farm, Agassiz, 1922.
ON LOAN FROM LORNE FISHER

Preparing land for hoeing, Dominion Experimental Farm, Agassiz, 1921.
FROM INTERIM REPORT FOR 1921 ON LOAN FROM LORNE FISHER

Select Swedish oats, Dominion Experimental Farm, Agassiz, 1921.
FROM INTERIM REPORT FOR 1921 ON LOAN FROM LORNE FISHER

Farm Employees, Dominion Experimental Farm, Agassiz, early 1920s. Back row, left to right: not identified, Mr. Lunt, Mr. Pennington, Alvin Ogilvie, C.B.A. Lovell, Nelson Hardy, Jim Fraser, not identified, not identified, G.C. Harper, Sam Stock, Isaac Duncan, Mr. Rancheau, not identified, Mr. Musselwhite. Centre row: Kenneth MacBean, Miss Archibald, Miss R. Keene, W.H. Hicks. Front row: not identified, not identified, Alex McKay, W.S. Moore, Vernon Kuhn, not identified.

AGASSIZ RESEARCH STATION:1886-1986

superintendent. By 1911, with the appointment of P.H. Moore as superintendent, the emphasis at the farm shifted from horticulture to livestock operations. The same year, the first Holstein-Friesians, consisting of 28 grade females and one purebred bull, were brought to Agassiz, the beginnings of what would become a world famous herd. W.H. Hicks took over as superintendent in 1916, a position he held for almost 40 years. In 1921, dairy research was identified as a major research emphasis of the farm and the grade herd was dispersed, leaving only registered animals.

The British Columbia Department of Agriculture

The provincial government recognized early the need to guide and support the development of agriculture in British Columbia. In 1873, an agriculture portfolio was drawn up under the Minister of Finance. In 1891, James Anderson was named Collector of Statistics for Agriculture under the Hon. J.H. Turner, then Minister of Agriculture. According to Anderson, Turner appointed him to organize his office upon "realising the importance of that branch of the public service which up to this time had been relegated to obscurity … On

Dominion Experimental Farm, Agassiz, 1939.

Harold Hicks

Harold Hicks

A man of great agricultural knowledge, W.H. Hicks was superintendent of Agassiz from 1916 to 1953. He was a leader in community and agricultural activities, and was BC's third Canadian Holstein Association president in 1947. In 1958, he was elected MP, a position he held until 1962. Mr Hicks died in 1974 at the age of 85.

BC Holsteins, 1886-1983.

Directors of the BC Dairymen's Association, 1910. Back row, left to right: James Turner, Royal Oak, Victoria; J.M. Steves, Steveston; J. Evans, Duncan; J. Thompson, Sardis. Front Row, left to right: W.H. Menzies, Pender Island; Wm. Duncan, Comox; F.J. Bishop, Victoria; J.E. Buckingham, Eburne; W.E. Scott, Department of Agriculture, Victoria; P.H. Moore, Dairy Commissioner, Victoria.
BC FARM MACHINERY MUSEUM

assuming office, I found absolutely nothing had been done, not the scratch of a pen, no books or papers to guide me. Without approbation, clerks or an office, the work was naturally arduous …"[4] The same year, Anderson wrote the first annual report on agriculture and outlined some of the problems that settlers were having, including "diseases and pests of fruit and vegetables, wild horses running at large in the ranching areas, blocks of uncultivated land locked up by private companies, lack of water for irrigation, and no creameries or cheese factories to buy farmers' milk."[5]

In 1894, recognizing that agriculture in the province had developed enough to require that specific attention be paid to the needs of farmers, the *Department of Agriculture Act* was passed to establish a Department of Agriculture to partner "with rural people to bring food to the table of British Columbia consumers. This meant finding ways to assist the new settlers to clear land, improve their crops and livestock, protect their harvest from pests and diseases, and help market their products."[6] Extension work was deemed important by the department even in these very early years, and it hired "a

C.T. 7

BRITISH COLUMBIA DEPARTMENT OF AGRICULTURE

DAIRY BRANCH—COW-TESTING (MILK-RECORD) DIVISION

No. N 532

Year 1937

Certified Milk and Butter-fat Record

CALVED WITHIN PERIOD

Dewdney--Deroche _____ COW-TESTING ASSOCIATION

Name of Cow Ruth Aggie Palestine DeKol   Registered No. if P.B. 159933   Breed Holstein

Date of Birth February 23, 1926 Age 9   Owner St. Mary's Residential School   Address Mission

2nd Owner _____ Address _____

Calved February 8, 1936   Calved after test February 12, 1937   Days in Milk 305   Average per cent. Fat 3.28

15,264   POUNDS OF MILK   501   POUNDS OF BUTTER-FAT

H C Clark
Supervisor

F C Morrish
Secretary of Association

E L Morrison
President

Date June 7, 1937

G. H. Thornber
For Dairy Branch

300-138-7165

Ear-Tag No.
DOT
DABCOP
Prov. T.B.
Dom. T.B.

Description of Cow

Black with white markings, white switch

Particulars of subsequent records on reverse side, also names of Pure-bred Sire and Dam (if known)   (OVER)

Milk and Butterfat Record of *Ruth Aggie Palestine DeKol*, **owned by Saint Mary's Residential School, Mission, 1937.**

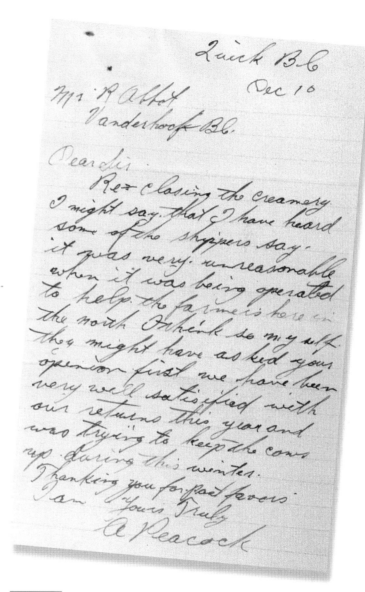

Quick B.C.
Dec 10

Mr R Abbot
Vanderhoof B.C.

Dear Sir

Re closing the creamery. I might say that I have heard some of the shippers say it was very unreasonable when it was being operated to help the farmers here in the north. I think so my wife they might have asked your opinion first we have been very well satisfied with our returns this year and was trying to keep the cows up during this winter. Thanking you for past favors.

I am Yours Truly

A Peacock

Letter from A. Peacock of Quick, B.C. to Mr. R. Abbott, Manager of Nechako Creamery, 1928.

Report by the Department of Agriculture on Progress in the Bulkley Valley

The industrial depression prevalent throughout the world is felt keenly here owing principally to the closing of the mines that are in the surrounding mountains, and to the shortage of orders for railway ties which usually gives employment to many men, but it has not affected the progressive dairymen very much. The houses may not be as luxurious as in some localities — although radios are numerous enough. The barns could be more up-to-date possibly, but the fact remains that these men are facing and solving the problems that confront them with common sense. They ask for advice from the right authority and they use the advice. When Prince Rupert enjoys the industrial expansion she expects and the natural increase in population which would follow, the Bulkley Valley dairyman intends to be ready to supply all the extra milk and cream that will be needed. He has pinned his faith on the cow, "the Mother of prosperity," and she won't fail him.

Dairying in the Bulkley Valley, 1931.

"Quality and Quantity": A picture taken at Dominion Experimental Farm, Agassiz, circa 1920. BCARS I 52281

number of lecturers to deliver papers on agricultural topics in all districts where farming could be said to have passed the pioneer stage."[7]

In 1908, the work of the Department of Agriculture was divided into three branches: livestock, horticulture, and poultry. The dairy division was added in 1912, the same year that cow testing associations were established. The department's work was province-wide and attempted to reflect on and respond to regional issues in a timely manner. For example, "early reports from the Fraser Valley indicated that milk production was highly seasonal. Too much milk was produced in the spring and summer months when there was an abundance of green grass, and too little in the winter months when there was a shortage of feed. The department began to promote silage making … and purchased a 6 hp engine and silage cutter with an elevator mounted on a truck, and offered to build and fill, free-of-charge, the first silo

VICTORIA
November 24th,
1 9 2 8

R. C. Abbott, Esq.,
Manager,
Nechaco Creamery,
Vanderhoof, B.C.

Dear Sir:

Your letters of November 17th and 19th, respectively,
came to hand on November 23rd and I had previously wired you
on November 22nd, as follows:-

"Close down creamery operations November thirtieth
prepare books for audit thereafter stop Salary to continue
to December thirty first stop Shippers will be officially
notified from this Department destination cream Valentin Dairy
Prince Rupert Price special thirty six Number one thirty four
Sutton will confer with you Letter following."

On the same date I wired the "Nechaco Chronicle" wording for a Notice to be inserted in its issues of November 24th
and December 1st and circular letter was forwarded to the cream
shippers shown on the list furnished to the Live Stock Commissioner by you under date of November 14th, a copy of which circular is enclosed herewith for your information.

You will, therefore, arrange to close down the plant
on November 30th and place same in safe condition for winter
as outlined in general on enclosed memorandum of instructions.
Will you kindly advise me at what date you anticipate it would
be possible for you to close out your patrons accounts for the
month of November and have your books prepared for audit.
This should be done as early in the month of December as would
be practicable with efficiency.

Mr. R. G. Sutton, District Agriculturist at Prince
George, will be in Vancouver during next week and general instructions will be given to him in writing and verbally as to
further procedure so that he may be placed in charge of the
plant during the period when it will be closed down.

In that interval a general policy will be determined

R. C. Abbott, Esq.   -2-

by the Government in relation to encouragement of the dairy
industry not only tributary to the Nechaco Creamery but in
other parts of the province. Until such policy is finally
determined it is impossible to make any further announcement
regarding the conditions under which the creamery will be
reopened in the Spring.

Mr. Sutton will be instructed to visit Vanderhoof
after his return from Vancouver some time after December 8th.
You will, therefore, kindly co-operate with him in carrying
out such instructions as he may have.

Yours very truly,

Wm. Atkinson

Minister

P.S.
Please notify butter customers as to date upon which you
will cease to make shipments of butter.

Letter from William Atkinson, Minister of Agriculture, to R.C. Abbott, Manager of Nechako Creamery, announcing the closure of the creamery, 1928.   BCARS ADD. MSS. 1150

It took me seven days of hard riding to reach Fort George to Telkwa. Now, with the advent of the railway, it is only an overnight trip.

Report of H.E. Walker, District Agriculturalist, Telkwa, 1913.

in each Farmers' Institute district. In the first year, five silos were erected and filled. This success led to the purchase of a second silo-filling unit and both units were placed under the direction of a silo demonstrator. In five years, 200 silos were in operation in the Fraser Valley and Vancouver Island. As predicted, the feeding of silage to dairy cows in the winter months further increased milk production."[8]

The department worked to improve the quantity of milk and butterfat produced in the province by assisting farmers to develop local cow testing associations, beginning in 1913. The department employed cow testers who travelled from farm to farm talking to farmers about their herds, taking a sample of milk from each cow, and testing the sample for its butterfat content. These comprehensive records on the production of each cow in a herd made it possible for farmers to make their decisions about selection not just on the basis of intuition, but in combination with fact. Regional branches of the Cow Testing Association were the precursors of the Dairy Herd Improvement Association (DHIA).

Other provincial initiatives were not as successful in the long term. In the early 1920s, for example, the department attempted to assist northern dairy farmers to secure markets for their milk by opening a creamery in Quesnel as well as the Nechako Valley Creamery in Vanderhoof. It was closed in 1928, amidst the protest of local farmers.

The department paid a good deal of attention to development in "new" areas of production. In a memorandum, likely dated 1929, surveying "the Economic Factors of Butter Fat Production in BC," H.R. Hare

DEPARTMENT OF AGRICULTURE

DAIRY BRANCH

Cream is now being purchased on a uniform grade or quality basis at all creameries in the three prairie provinces. There are four recognized grades, namely "Special," "First," "Second," "Off Grade." A premium is paid for the higher grades and every creamery manager prefers to purchase these grades at the higher price.

Observe *Cleanliness* and *Coolness* in handling. These are the keys to quality cream.

If you are receiving "No. 2," or "Off" grade on your cream, consideration of the following explanations should assit in remedying the difficulty.

Common Causes of Cream Being Graded No. 2

HIGH ACID—
Cream not cooled quickly after separating, or held at too high a temperature. *Cool immediately to 40° F., or as near as possible, and hold at this temperature.*

STALE OR BITTER—
Cream held too long. Ship often. If cans are too large, it will pay to get smaller ones.

METALLIC—
Rusty cans, milk pails, or separators.

HEATED—
Mixing cream from different milkings without being properly cooled and stirred before mixing. Mix all cream thoroughly whenever a fresh lot is added to the shipping can.

WEEDY—
Do not allow cows on pastures containing Wild Onions, Frenchweed, Rag Weed, Wild Mustard or Willow brush.

UNCLEAN—
Wash thoroughly milk pails and separator immediately after use. Rinse with boiling water. Use a brush only in washing utensils, avoid dish cloths. See that cows' udders are clean. Milk with dry hands. Avoid old dirty corrals.

VEGETABLE OR CELLAR FLAVORS—
Avoid badly ventilated, musty or mouldy cellars or cream being exposed to decaying vegetables or fruit.

Causes of Cream Being Off Grade

YEASTY—
Contains the germs of yeast obtained from unclean utensils combined with exposure to *high temperatures*.

OILY—
Fumes from coal oil lamps or gasoline. Do not put coal oil or gasoline in cream cans. Use care in touching oil or gasoline or working with gasoline engines when handling dairy products.

STINKWEED or FRENCHWEED, ONIONS, GARLIC—
Keep cows away from these.

Information circular, Dairy Branch, Department of Agriculture, circa 1920.
BCARS GR 509

34

Moricetown BC
15th August 27

The Manager
The Nechaco Creamery
Vanderhoof B.C.

Dear Sir,
We received your letter with July Statement
and Circular Letter. We are sorry to hear there is a
possibility of closing down during the coming winter
as we were hoping to be able to ship some cream
to you right along and reckoning that the creamery
would be going all winter we bred the cows
early and they are due to freshen about 18th Jan 2nd Feb
3rd 16th & 23rd March. With this end in view we have
planned out operations for this season and we rely
on the cream for the main revenue. We realize
however that as we are only in a small way our
shipments would not carry much weight in any
decision that may be arrived at.
In order to try & boost your sales somewhat
we spoke to a local storekeeper — Mr W. H. Maclean —
and have pleasure in asking you to send her a box
of butter in 1 lb prints. We do not know if you

handle it in less than 50 lb lot but if you
can send 25 lbs failing this send the 50. You
may debit the amount to our A/c.
Yours faithfully
A, T & G. Campbell
P.O.B.C.

Ship Butter to.

Mr W H Maclean
Moricetown BC.
express prepaid as there is no agent

Letter from A., T., and G. Campbell, Moricetown, to the
Manager of the Nechako Creamery, expressing their
concerns about a possible plant closure, 1927.
BCARS ADD. MSS. 1150

Herdsman at Colony Farm

I started at Colony Farm in 1949. I was raised on a farm in Prince Edward Island and served in the armed forces during the Second World War. When I came back, I wanted to be a herdsman with dairy cattle. I helped my father for a few months in the summer trying to make up my mind what I was going to do. I came to BC because BC paid the highest wages.

I was hired as a milker at Colony Farm. Later, I was hired as a herdsman and was responsible for taking care of the herd — the milking and the breeding and everything involved with the dairy herd. At Colony Farm, they had pigs and sheep and fruit and vegetables, and so there was somebody in charge of every section. The whole staff was around 50 — that was the field crew, the crew for the piggery, the sheep, and the office staff.

In the old days, there were mental hospitals all across Canada. Every mental hospital had a farm and they had two purposes for the farm — to provide good food for the patients and also to provide some type of work and exercise for them. The patients used to do a lot of work. The farm didn't supply all of the food to the hospital, but it supplied a lot of it. We used to slaughter about 30 pigs a month and that pork all came into the hospital. The dairy had a complete processing plant. The milk was pasteurized and bottled in various sizes for the hospital and cream was provided. They didn't do butter or cheese or any other dairy products. Just milk and cream. They grew a lot of potatoes and vegetables. All the bull calves were raised for veal and that veal came back to the hospital.

Some patients were good enough to be working on their own at certain jobs. Other patients were put into gangs with a nurse from the hospital in charge. They did a lot of work — they shoveled snow in the winter time, they picked the potatoes and vegetables, they helped stack the hay. I had some real good patients over the years that worked in the barns.

The entire place was more or less burnt down in 1947 — called the "big fire." It burnt the big arena. They lost the horses. Previous to that there were lots of other fires. And there were other fires in later years. In 1975 or 1976, there was a fire in the calf barn. All the calves were in pens — about eight to a pen. When I saw that the fire was starting, I went along and opened the gates and kicked them out of the barn. At one barn with Dutch doors, the top was open and the bottom was closed. When I went to reach over the top of the door to unlatch it, the top was so hot that I couldn't. I kicked the door in. I got the calves out and I pulled the door shut, but I couldn't latch it. All those calves went back in and they all burned to death.

The most animals — cows, calves and bulls — I had to look after at one time was 555 head. It scaled down a little bit from that. We milked as many as 250 cows. It would fluctuate a bit over the years.

Colony Farm was closed in 1983. I hated to see a life's work dispersed like that. That's 17 years ago. I guess that there isn't an animal living that's got the Colony prefix.

Bill Howe, 2000

makes special note of the area from Quesnel north to Prince George and of the valley west of Prince George. "This is the newest dairying area in BC, and has the greatest possibilities," he writes, because "a very small proportion of the farm land has been cleared and farms are quite scattered."[9] But the area was not without difficulties, he notes, for while "a market for a limited quantity of farm products is provided by the small cities of Prince Rupert and Prince George and a few towns, the two creameries that are operating within the area find it difficult to assemble economically."[10] As well, the area has long winters and a short frost-free season, so "the marketing of butter is difficult. The butter produced during the summer season meets competition from the trade that necessarily imports from other sections during the winter."[11] Many farmers, then, found it necessary to supplement their income with work off the farm, by working in the woods in the winter, or by finding some other source of revenue.

Colony Farm

In 1910, the dairy herd at Colony Farm was started on the banks of the Fraser River in conjunction with the establishment of the provincial psychiatric hospital at Essondale. While the early herd began with 70 randomly selected females, "only a few of this original group lived to make a lasting imprint on the herd."[12] P.H. Moore was appointed superintendent in 1916, and "he embarked on a line breeding program within the herd that stressed high production and approved type" and as one commentator stated, "those that didn't make history, made hamburger."[13] Moore's attention to herd

One of the original buildings at Colony Farm, used for patient housing, offices, and
staff cafeteria.

Horse arena with dorms and offices upstairs, Colony
Farm.

Barns at Colony Farm.

Entrance to Colony Farm.

Three foundation cows purchased in 1912 by Dominion
Experimental Farm, Agassiz, from J.M Steves.

Colony Farm

For many years, the names of Colony Farm and Pete Moore were synonymous in the public mind. Colony was one of Western Canada's claims to distinction and the man behind it was Percy Homer Moore, universally known as "Pete."

Grant MacEwan, 1952

Colony Farm and Essondale, 1920s.

BCARS G 04642

development meant that after 1920 very few bulls and no new females were brought into the herd. "For a period of thirty years, only four bulls were bought. Moore's line breeding program worked, and some of the most outstanding Holsteins in North America were developed at Colony Farm."[14]

Some important sires of this period were *Hazelwood Heilo Sir Bessie* and *Sir Romeo Mildred Colantha* – the first two Century Sires of Canada. Important dams were *Colony Fleta Heilo* and *Colony Flood Colantha*, All-Time All-American Three-Year-Old. Pete Moore's hard work and the exceptional quality of the Colony Farm herd were recognized when it received the Holstein-Friesian Asso-

ciation of Canada's first Master Breeder award in 1930.

Art Hay took over as farm superintendent after Moore's retirement in 1948 for a period of four years. At that time, W.B. Richardson took over the helm, with Bill Howe as herdsman, and continued to build on the strong tradition of excellence. The Colony Farm herd was dispersed in 1983.

Farming Associations

There are many associations, some of which continue to the present day, that supported the building of a professional community of dairy farmers. The BC Dairymen's Association was chartered in 1894 and worked to bring

38

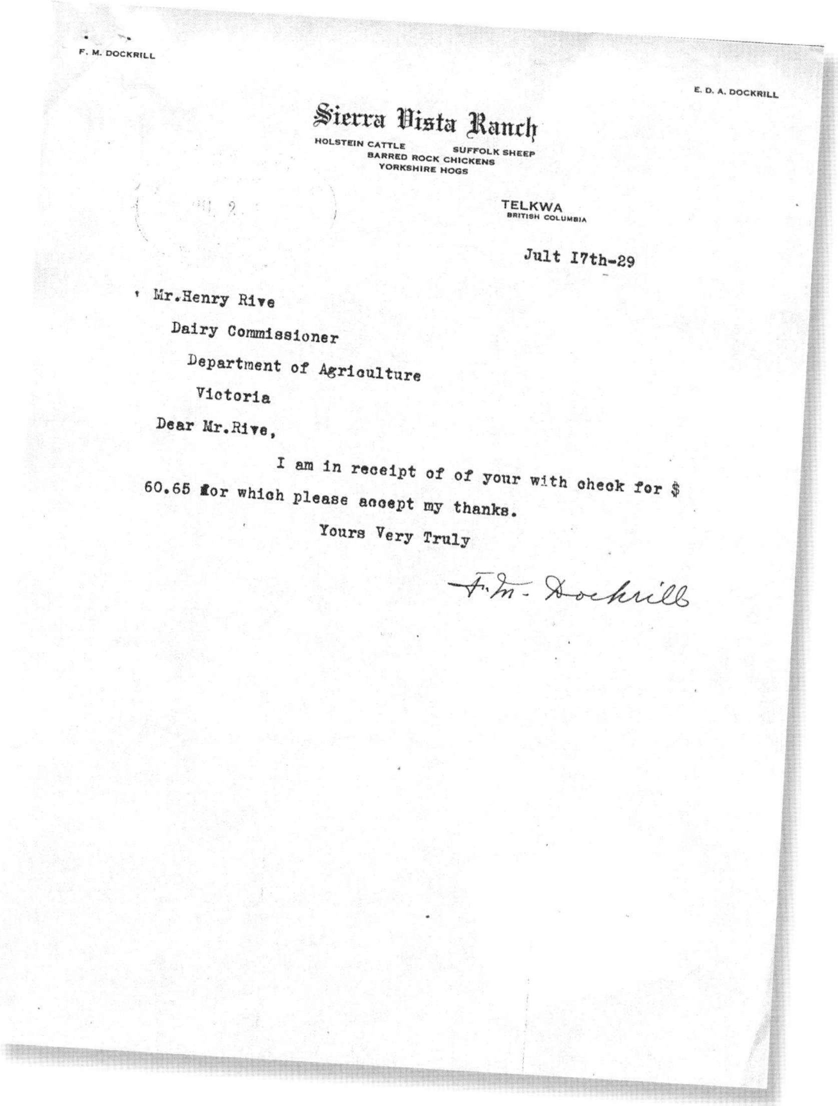

F. M. DOCKRILL   E. D. A. DOCKRILL

Sierra Vista Ranch

HOLSTEIN CATTLE   SUFFOLK SHEEP
BARRED ROCK CHICKENS
YORKSHIRE HOGS

TELKWA
BRITISH COLUMBIA

Jult 17th-29

Mr. Henry Rive

Dairy Commissioner

Department of Agriculture

Victoria

Dear Mr. Rive,

      I am in receipt of of your with check for $
60.65 for which please accept my thanks.

      Yours Very Truly

          F. M. Dockrill

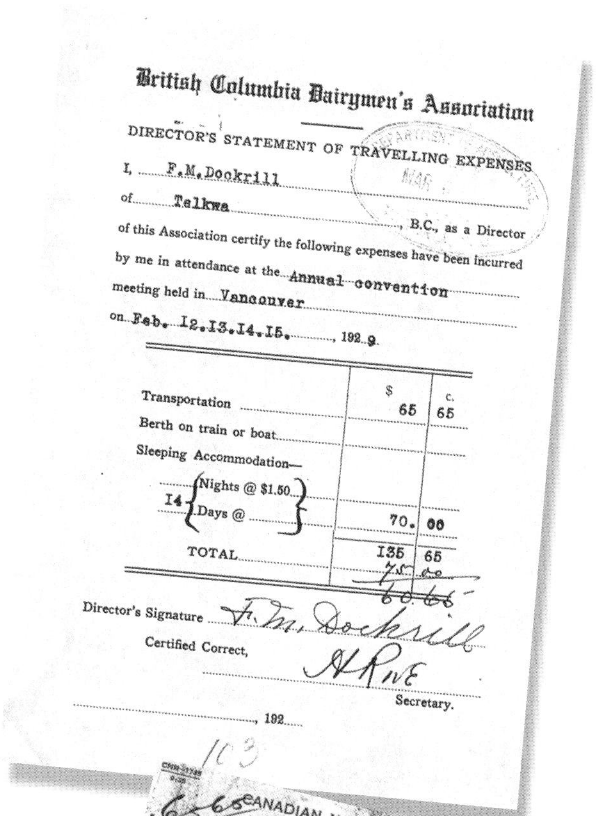

British Columbia Dairymen's Association

DIRECTOR'S STATEMENT OF TRAVELLING EXPENSES

I,   F.M.Dockrill

of   Telkwa   , B.C., as a Director

of this Association certify the following expenses have been incurred

by me in attendance at the Annual convention

meeting held in Vancouver

on Feb. 12.13.14.15. , 192 9.

	$	C.
Transportation	65	65
Berth on train or boat		
Sleeping Accommodation—		
14 { Nights @ $1.50		
{ Days @	70.	00
TOTAL	135	65
	75.00	
	60.65	

Director's Signature   *F. M. Dockrill*

Certified Correct,   *H Rive*

                                Secretary.

                , 192

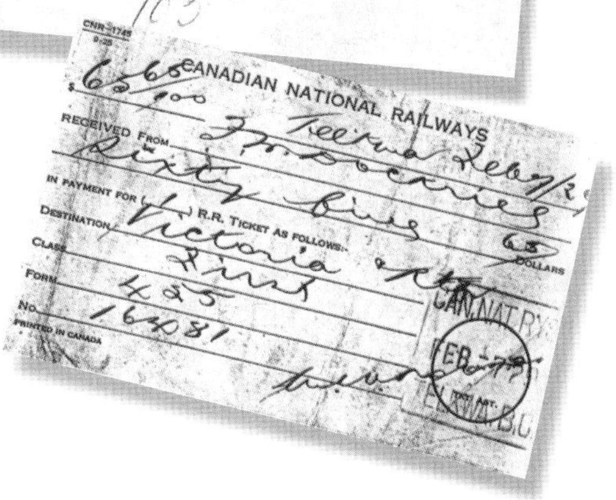

CANADIAN NATIONAL RAILWAYS

$ 65/100   103

RECEIVED FROM

IN PAYMENT FOR

DESTINATION Victoria

CLASS First

FORM 425

NO. 16481

PRINTED IN CANADA

Expense claim to BC Dairymen's Association, F.M. Dockrill, Telkwa, 1929.

Director's Report to the BC Dairymen's Association, 1908 or 1909

It gives us great pleasure to report that for the last year, there has been a steady increase in the dairy products manufactured in the province, that milk and cream have been delivered to the factories in improved condition, and the market returns for dairy products have greatly increased over the previous year. While there has been an advance in price, owing to the better and more uniform quality of butter manufactured, yet we urge the dairymen to still greater effort, as there is yet much room for improvement in the methods of handling and caring for the milk or cream before it is delivered to the factories. We would also like to point out that many of the butter makers might be more painstaking in keeping their premises in a neat, sanitary condition, thus setting a good example to their patrons. We would urge the dairymen of the province to weigh and test each cow's milk, keep a record of every cow's performance, so that unprofitable animals may be weeded out, when they may be replaced by animals which will give a profit of the food consumed. By following such a system the profits and produce of a herd may be doubled.

Delegates, BC Dairymen's Association annual convention, Victoria, 1910.   BC FARM MACHINERY MUSEUM

the concerns of milk producers to government. According to its constitution, "the object of the Association shall be to encourage the improvement of the milk, butter, and cheese industries in the province of British Columbia."[15] The Dairymen's Association sponsored a range of speakers on topics pertinent to dairy farmers. In 1895, at one of the very earliest meetings, it was noted that "Mr. R.H. Caswell read a paper on care and handling of milk and exhibited a Babcock Tester."[16] At the association's annual general meeting in 1908, Dr. Knight, "veterinary surgeon and inspector for the Dairymen's Association," spoke about his inspection of 400 farms and dairies in the province. According to the minutes, "his chief theme was the spread of Tuberculosis all over the country, and what might be done to eradicate or alleviate this disease including the enforcing of cleanliness in the dairy and stable, plenty of space, sunlight and fresh air."[17] The Director's Report of the same

Expense claim to BC Dairymen's Association to attend Farmers' Advisory Board Meeting, J.W. Berry, 1928. BCARS GR 510

J.W. Berry. Expense claim, 1928.
BCARS GR 510

Minutes of first meeting of BC Dairymen's Association of BC, 1894. BCARS GR 510

Elmore T. Paull Remembers

Shortly after we moved to Agassiz, I got a job milking test cows four times a day at the Agassiz Experimental Farm, where I worked for three years. In 1927, Superintendent Harold Hicks suggested that we take our herd to the shows and offered me time off to help show them so about one week before the Vancouver Fair I went home to help. At that time, we had 21 head and we took 14 to the fairs. We showed at Vancouver, then Victoria, then back to the Chilliwack Fair, then New Westminster, then home. At Vancouver, we met Art and Al Hay who were showing the D.A. McPhee herd from Vankleek Hill, Ontario. McPhee's herd sire, *Sir Francy Mercena Burke*, had been All American Three-Year-Old the year before. The Experimental Farm was showing *Tsussie Rajah*, a bull they had purchased the year before in a BC Holstein Breeder's sale from Henry Bonsall and Sons of Westholme, one of the pioneer herds on Vancouver Island. There was quite a lot of betting the night before the show as to which bull would win, but *Sir Francy* was the winner. I do not remember who the judge was, and I'm still not sure if the right bull won. They could have gone either way. *Tsussie Rajah* proved to be only a fair breeding bull, but was a nice bull to handle. Horatio Webb of Sardis, an elderly man, rode him in the stock parade. He had won several grand championships for the Bonsalls before Agassiz bought him. When they were fitting him for the shows, they used to give him a couple of buckets of milk, night and morning. I don't know how much he would have drank if it had been given him. It really put the bloom on him. His hide was just like silk.

On the night before the Vancouver show, while we were watching the lines, Art and Al Hay, and my father and I talked almost all night. They were showing a son of *Francy* named *Texal Burke of Crystal Spring* as a junior yearling. His dam was a daughter of a then Canadian champion. We made a deal to buy him after the Victoria Fair, as they were going south from there to Puyallup and Spokane. He was Junior Champion at Vancouver and Victoria, and then Junior Champion for us at Chilliwack and New Westminster. There was lots of competition. Besides the McPhee herd, there were Colony Farm, Agassiz Experimental Farm, Frasea Farms, Harry Brown of Surrey, Bill Woods from Surrey, J.H. Mufford and Sons from Milner, and J.W. Berry and Sons of Langley. At Victoria, Ralph Rendle was showing some and Raper Brothers, also of Victoria, had a small string. They both retailed their own milk. At Chilliwack, A.W. Annis and Son had some nice cattle and Hogg Brothers (Reuben and Clarence) of Agassiz had a few. Most of the herds that showed at Vancouver were at New Westminster, and the Laing Herd from Sea Island was also there. Another important Sea Island herd was that of Captain Erskine and Son. Later they moved to Ladner where they bred some excellent cattle.

BC Holsteins, 1886–1983.

year urged farmers to work to control this disease because "the checking, or eradicating, of bovine tuberculosis is one of the greatest problems that the dairymen of BC have had to face. We can not close our eyes, or think lightly of it. Each year sees a wider spread of the disease, causing a greater loss to the breeders and claiming more victims among our fellow men. Therefore we earnestly ask that each dairyman have his cattle tested and clean the dread disease out of the herd before it is too late, and the herd, which has taken years of patient toil and care to build up, becomes a loss."[18]

Like the Dairymen's Association, the Farmers' Institute was supported by the Department of Agriculture as a way to represent farmers and their interests to government and to provide extension education to farmers in British Columbia. In 1897, an organizer from Ontario came west to help set up the first chapters of the Farmers' Institute; the Surrey-Langley Farmers' Institute and the Richmond Farmers' Institute were both chartered before the year was out. By 1909, there were 8,000 members in the province participating in meetings, shows, short courses, lectures, and research. By 1912, 77 institutes had been formed. The Department of Agriculture was pro-active when it came to signing farmers up, as the following excerpt from W.E. Walker's Annual Report of 1917 suggests. He complains that "to organize Farmers' Institutes in the Cariboo, we travelled by boat 110 miles down the Fraser River, and 250 miles by model 'T', horseback and buggy — some in snow 14 inches deep, only to find when we arrived in Clinton — not a single farmer showed up for the meeting. Successful meetings were held at Quesnel, Soda Creek, 150

1913

Annual Convention

OF THE

B.C. Dairymen's Association

TO BE HELD

Thursday and Friday, January 30 and 31

IN THE

City Hall, New Westminster, B.C.

PRESIDENT: F. J. BISHOP, DUNCAN

VICE-PRESIDENT: JOS. THOMPSON, SARDIS

SECRETARY-TREASURER: H. RIVE, DEPARTMENT OF AGRICULTURE, VICTORIA

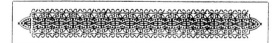

Program of speakers, BC Dairymen's Association convention, 1913.

BCARS GR 510

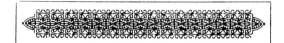

Programme

THURSDAY, JANUARY 30TH.

9.30 A.M.

Opening address : HIS WORSHIP THE MAYOR.

Address : F. J. BISHOP, President.

Business meeting : election of officers.

11.00 A.M.

Address : "The Dairy Sire."—HUGH VAN PELT, formerly Iowa State Dairy Expert.

2.30 P.M.

Address : "Forage Crops."—P. H. MOORE, Supt. Dominion Experimental Farm, Agassiz.

3.30 P.M.

Address : "Improving the Dairy Herd."—F. H. SCRIBNER, U.S. Bureau of Animal Industry, Wisconsin, U.S.A.

7.30 P.M.

Address : "Utilizing the By-products of the Dairy."—J. R. TERRY, Provincial Poultry Instructor.

8.30 P.M.

Popular illustrated address : "Dairy Inspection." —L. W. HANSON, Deputy Dairy and Food Commissioner, Seattle, Wash.

Programme

FRIDAY, JANUARY 31ST.

9.30 A.M.

Address : "The Management of the Dairy Cow." —HUGH VAN PELT.

11.00 A.M.

Address : "Business Methods for the Dairy Farmer."—THOS. CUNNINGHAM, Mgr. Farm Dept., Western Fuel Co., Nanaimo.

2.30 P.M.

Address : "Community Breeding."—W. T. McDONALD, Provincial Live-stock Commissioner.

Presentation of Dairy Farm Competition Trophies and Medals and B.C. Record of Performance Cups.

4.00 P.M.

Address : "Cow-testing Associations." — F. H. SCRIBNER.

7.30 P.M.

Address : "Milk and its Products in Relation to Health."—DR. A. P. PROCTOR, Vancouver, Chairman Royal Milk Commission.

8.30 P.M.

Business resumed.

9.30 P.M.

Informal Banquet.

The HON. PRICE ELLISON, Minister of Agriculture, and MR. W. E. SCOTT, Deputy Minister, are expected to attend and give addresses during the Convention.

Mile, 100 Mile and Ashcroft."[19]

The Department of Agriculture further supported development of farmland by buying stumping powder by the railcar load and then selling it at cost to Farmers' Institute members. Indeed, in some areas of the province, after veterans had returned home from the First World War and taken up land, residents noted that land clearing with stumping powder was proceeding at such a pace that it sounded as if the war had been transplanted to Canadian soil.

The Women's Institute was also supported by the provincial Department of Agriculture as a way of developing an agricultural network in the province and making sure that farm families had the information and the social and community support structure to succeed, especially in remote areas. Introduced to British Columbia in 1909, the WI spread rapidly: by 1910, 30 branches had been formed and approximately 1,300 members had joined. A provincial Women's Institute Advisory Board was formed to give women a voice in Victoria, to promote better education and support for rural families, and to create workable health services for rural areas. The Women's Institute Report, contained in the 1913 Annual Report of the Department of Agriculture, lists speakers and demonstrators available to give lectures to Women's Institute branches. Among others, Mrs. S. Davies, a graduate of Warwick Agricultural College for Ladies, residing at 210 Grant Street in Vancouver, was willing to give lectures on "poultry, dairying, floriculture, and market gardening." Her peer, Miss Gerrard of Royal Oak, would travel to lecture on "Dairying for Women" and "The Travelling Dairy."[20]

Hazelmere Women's Institute Declaration of Association, circa 1911. LANGLEY CENTENNIAL MUSEUM AND NATIONAL EXHIBITION CENTRE

USE CXL STUMPING POWDER
TO OBTAIN THE BEST RESULTS

Ask the Secretary of the Institute in
your district for full information or
write

Canadian Explosives, Ltd.

Room 913 Birks Building

VANCOUVER, B. C.

Stumping Powper Booklet Mailed Free on
Request

Advertisement for stumping powder sold by Farmers Institutes.

Vancouver, B. C.,
June, 1918.

To Our Churning Cream Shippers.

Gentlemen:

With a view to effecting improvement in the quality of our B. C.
Creamery Butter, we find it advisable to adopt the modern system of
paying for cream on a QUALITY BASIS.

At the present time 90% of the churning cream has been
received here in an overripe condition.

We have decided to adopt the system of CREAM GRADING, which
has proved so satisfactory to both patron and creamery in Alberta,
Saskatchewan and Manitoba, and is approved by the B. C. Department of Agriculture, and in future we will credit our shippers
as follows:-

FOR FIRST GRADE OF CREAM testing not less than 25% and not
over 40%, acidity very mild and free from undesirable flavors, we
will pay a BASE PRICE, in accordance with the market price of butter.

FOR SPECIAL GRADE CREAM testing not less than 25% and not over
40%, sweet or **practically sweet**, clean and free from undesirable
flavors, we will pay 2 cents premium above the BASE PRICE.

FOR SECOND GRADE CREAM classed as being overripe or strong
flavored or too low in test we will pay 2 cents less than the BASE
PRICE.

Any cream received in a condition unfit for the manufacture
of creamery butter will be returned at shipper's expense.

We trust you will be encouraged by the premium offered to
use every precaution possible to keep your cream clean, cool and
sweet and by so doing co-operate with us in raising the reputation
and value of our local product, in competition with butter imported
from other sources.

Yours faithfully,

A. P. SLADE & CO.
VANCOUVER CREAMERY CO., LTD.
DAVID SPENCER CREAMERY DEPT.
FRASER VALLEY DAIRY.
P. BURNS & CO.

**Vancouver area creameries adopt system of paying for cream on
quality basis, 1918.**

We Are Not Milking Cows for Honour and Glory — What For Then?

Do we milk cows because we love them or to keep ourselves busy or for the wrist and finger strengthening exercises? To keep the creameries running or for our own profit? Presumably for profit.

J. Rive, BC Department of Agriculture, 1939 or 1940.

Delegates to Western Canada Dairy Convention, sponsored by the Department of Agriculture, 1929.

Western Canada Dairy Convention

H. J. Hartshorn – Huntingdon. B.C.
W. Wood   R.R.1 Cloverdale B.C.
C. Erikson   R.R.3 Cloverdale B.C.
W. Newton   Vancouver B.C.
O. July   Steveston   B.C.
Thomas Sooter   Glen ____
W. Scott Walker Barnston Island B.C.
Harry Bay   R.R.3 Abbotsford
G.R. Addison   Ft. Langley
G.A. Medd   Ft. Langley
A. Sharpe   Chilliwack B.C.
D.S. Heelas   Vancouver
J.J. Edwards   Chilliwack
W. Rose   Langley Prairie
G.W. Guerin   Chilliwack
W.J. Marlin   Westminster
A. Scott   Mayfair Butter Co   Vancouver
A.J. McKean   Asst Gen Frt Agt CPR Vancouver
A.E. Steves   Steveston
J.M. Steves   Steveston
Robert W. Hornby Armstrong B.C.
____ Logan   Vancouver B.C.
J.B. Page   Ngligit
J.B. Watson   Sardis
H.E. Jackson   Atchelitz
____ Young   Ryewood
Stanley J. Borland Vancouver B.C.

J.M. Shaw   Westminster ____ Co
R.F. Arnett   New Westminster ____
J. Johnson   Co. Smith ____
J.J. Ward   Milner B.C.
Wm. ____   Milner B.C.
Wm. Dungate   Valentine Dairy Prince Rupert

Residents of B.C.
C. Fice   Victoria
James Aitken   Cloverdale
F. Bose   Surrey Centre
F.H. Norton   Victoria B.C.
S.H. Shannon   Cloverdale BC RR
By G. Harrison   Langley Prairie
H.J. Smithson   Vancouver
____ Colman
Edwin A. Wells   Sardis, B.C.
Richard Farquhar R.R. Cloverdale
B.W. Rushton   Langley Prairie
Wm. Scott R.R.1 Cloverdale
R.H. Moore   Essondale
A. Brehaut   New Westminster
A.H. Jagger   1016-41/3 Van. B.C.
Wm. Lord   Aldergrove B.C.
A.W. Drake   Ganges B.C.
J. Robertson   Vancouver B.C.
W.J. Park   Van   BC
S.H. Gilmore   Lulu Is
John Savage   Matzqui Prairie Dewdney P.O.
W. Jno Egan   2016-43 W. Vancouver
T.E. Kerr   Chilliwack B.C.
J.A. Bunthy   Royal Oak P.O. V.I. B.C.
Edward Romano   66-11th East Vancouver
J.B. Munro   Victoria
J.S. Dickenson   Angains B.C.
G.D. B. Hall   Chilliwack B.C.
Jas Bailey   Sardis B.C.
H. Dalzie   Dewman Fold
J. Moore   Vancouver
A.J. Williamson Victoria Vimpa Jersey Milk
E.W. White   Victoria
F.W. Clarke   Vancouver

Western Canada Dairy Convention
G. Grossman   U.B.C.   Chilliwack
Jas. Douglas   U.B.C.   Victoria
Wm. Hinsworth   U.B.C.
Peter Grossman   U.B.C.   Chilliwack
Robert A. Hornby O.B.C.
J. Ellis   U.B.C.   Armstrong B.C.
A.C. Richards   Exp. Farm   Vancouver
J. Freeman   Agassiz B.C.
Ernest F. Sherwood   C. Clayburn B.C.
J.S. Dunn   Vancouver
                             Nanaimo

Fourteen Fords

Whereas in earlier times, officials of the Department of Agriculture travelling from place to place in the course of carrying out their duties had to rely upon rail transportation and the horse and buggy, many were now making use of motor vehicles. By 1920, the department was operating "fourteen Fords and one Overland." In that year, these were driven a total of some 61,000 miles at an average overall cost of just over seven cents per mile.

*M.M. Gilchrist.
"A History of the
British Columbia
Department of
Agriculture," 1966.*

OREGON STATE AGRICULTURAL COLLEGE

SCHOOL OF AGRICULTURE AND EXPERIMENT STATION

CORVALLIS

February 15, 1929

Mr. Henry Rive
Department of Agriculture
Victoria, British Columbia

Dear Mr. Rive:

I arrived home at midnight last night and was immediately told by the good wife that I was due at a hearing before the Ways and Means Committee of our legislature tonight. I have just had time to get my material together and within a few minutes will leave for the state capitol. While I regred not being able to stay over I am glad that I came home when I did.

I want to say to you that I enjoyed my brief sojourn very much. I have attended a great many conventions of different kinds of dairy farmers and manufacturing men here in the states and outside of the purely scientific meetings I have never seen interest displayed equal to that which I have seen on my two visits to British Columbia. While I spent a number of hours of my own time and used the time of some of the assistants here in the department in getting this material together in order to present to you I feel amply repaid for my efforts.

I enclose my expense account which has amounted to more then I anticipated in the beginning. I have just secured a receipt from the local railroad station which I think will take care of the transportation affair satisfactorily and there are Pullman and hotel receipts also attached.

With best personal regards and wishes for a successful year's work, I remain

Yours very truly,

P. M. Brandt,
Professor of Dairy Husbandry

PMB:MP
Enc.

Expenses of P. M. Brandt incurred in travelling to Vancouver B. C. to address Western Canada Dairy Convention,

February 13, 1929.

Feb. 12	Railroad fare from Corvallis, Oregon to Vancouver, B. C. and return.	28.14
	Taxi in Portland	.55
	Parlor, Portland to Seattle	1.50
	Dinner on train	1.25
Feb. 13	Sleeper from Seattle to Vancouver	3.00
	Taxi at Vancouver	.65
	Breakfast at Vancouver	.95
Feb. 14	Breakfast at Vancouver	.75
	Hotel at Vancouver	.65
	Luncheon at Seattle	4.60
	Dinner at Portland	.50
		.85

$ 42.99

FEE 30.00

72.99

#75.00

Letter from P.M. Brandt to Henry Rive, Department of Agriculture, regarding his speaking engagement at the Western Canada Dairy Convention in Vancouver, 1929. BCARS GR 510

48

P.M. Brandt's bill at the Hotel Vancouver, 1929. Note the price of rooms.　BCARS GR 510

Rail Tickets, P.M. Brandt, 1929.　BCARS GR 510

Rail Ticket from Portland to Vancouver, P.M. Brandt, 1929.　BCARS GR 510

Holstein Herd At UBC

University of British Columbia, home of one of Canada's best Ayrshire herds, has now been given the foundation of a Holstein herd through the generosity of breeders in the province. The presentation of 32 heifers was made on May 4th at the annual field day of the BC Holstein Branch.

W.H. Hicks, secretary-fieldman, who had been responsible, along with Dr. J.C. Berry, for the selection of the heifers, made known the list of donors: Ten were from Colony Farm, three from J. Grauer & Sons, two from Les Gilmore and one each from Arrowsmith Farms, Archie Barker, Harry S. Berry, Wm. C. Blair, Harry Bose & Sons, Robt. W. Bridge, H. Leslie Davis, Agassiz Experimental Farm, Giacomazzi Bros., Houston Estate, Henry T. Jensen, Tony Lagemaat, Frank W. Machell, S.J. McKimmon, North Star Farms, Ralph & Gordon Rendle and James Threlfall. All but five of the group are bred heifers and all breeders present were impressed by the quality of the donations.

The Holstein-Friesian Journal,
June 1956.

The University of British Columbia

By Al Tuchsherer

In the early 1900s, many in the agricultural community believed that BC needed a school to work with the Experimental Stations to give people an opportunity to become more knowledgeable about the science and practice of agriculture. This thought became a reality in 1914 with the appointment of Dr. Leonard Klinck as the University of British Columbia's first Dean of Agriculture. It was Dean Klinck's responsibility to build a faculty that would reflect, support, and enhance British Columbia's diverse agricultural activities. In 1918 Professor Wilfred Sadler left McGill University to establish the Department of Dairying at the University of British Columbia.

An early priority of the department was the development of extension programs and short courses. Extension courses were set up as field days for farmers in rural parts of the province. Organized and taught by faculty members, short courses were several days to several weeks in length and were usually held at the university in the winter months. When the constraints of the Depression and high travel costs made it difficult for the University to offer rural field days, an extension initiative in the form of weekly radio lectures was introduced. These radio lectures, produced in partnership with the agricultural division of BC Electric, were available as Friday night bulletins to "forward thinking farmers who wished to make use of the information."[21]

Research expanded throughout the 1920s and into the 1930s under the guidance of Dean Frederick M. Clement. In 1928, Dean Clement served as chairman of a royal commission on milk, an investigation into BC's combative dairy industry. Professor Sadler and Dr. N.S. Golding studied the development of varieties of cheese suited to the resources of BC's rural farmers.

Beginning in the 1930s, Blyth Eagles assumed responsibility for the training of students in the Dairy Department. Eagles graduated from the UBC's Department of Agriculture in 1922, became a professor in 1929, and became Dean of agriculture in 1949. Eagle's research focused on the quality control of milk and milk products and the mineral content and nutritional value of BC milk.

The 1950s marked a period of technological and scientific advancement for the dairy industry. At this time the dairy short course was adapted and developed to meet the changing demands of the province's dairy industry. Topics included pasteurization, quality tests for milk, plant practices, bacteriology, chemistry, Babcock testing, ice cream making, and applied dairy bacteriology. In 1950, Dr. J.R. Campbell studied the quality grading of milk and designed a procedure that was later to be employed by the province's new *Milk Act*.

Over the second half of the century, agricultural studies at UBC have shifted focus. The university farm on the Point Grey site was eliminated and the herd transferred to the Agassiz Experimental Station. Programs have become more degree oriented. Today, the Faculty of Agricultural Sciences "focuses its teachings and research on food security and land use management. Its core values are grounded in the concept of sustainability — balancing ecology, economy and community to provide for ... the care and conservation of plants, animals, soil and water."[22] ■

Unloading the first Ayrshire herd, port of Vancouver, 1929. This herd came to UBC from Scotland. To allow broader research and to give students exposure to different breeds, Holsteins and Jerseys were later added to the UBC herd.

UNIVERSITY OF BRITISH COLUMBIA SPECIAL COLLECTIONS

Deans of Agriculture, left to right, the new dean Michael Shaw, early dean Dr. F.M. Clement, former dean Dr. L.S. Klinck, also former president of the university, and retiring dean Blythe Eagles, 1967.

DAIRYWORLD COLLECTION

Extension courses in rural BC and short courses at the university were considered an important service. Here, a lecture is underway, circa 1930s.

UNIVERSITY OF BRITISH COLUMBIA SPECIAL COLLECTIONS

CHAPTER III

Milk Cooperatives

Salesmen balancing their accounts in the checking-in room at Dairyland's 8th Avenue plant, 1948. Drivers on left: D.R. Setter, P.L. Bergh, E.R. Evans, J. Thomson, F. Brooks, J. Gordon, P. Humphries, J.T. Gavin. On right: A. Turley, D.H. Whiteford, G.T. Blackman, E.G. Meade, O.F. Tuffer, J.P. Davis, S.L Gray (Inspector), G.G. England. Standing: W.C. Pownall, cashier; F.B. Durrant, driver; J. Stothard, driver; T.G. Kennedy, driver; J.A. Carson, route manager; H.C. McBride, sales desk.

DAIRYWORLD COLLECTION

NOCA Memories

By J. Saunders of Vernon, 1965

I remember, I remember,
How forty years ago,
The dairy farmers at that time,
Were wondering where to go.
For where to sell their butter then,
They really did not know.

I remember, I remember,
Those grim days of old,
When poverty stood facing us,
And hearts felt pretty cold.
When NOCA's little group, just formed
Gave us a hand to hold.

I remember, I remember,
The cream cans on the rack,
And some did hold a hundred pounds
To strain a poor man's back,
And how poor Nelson had to do
The work of power jack.

I remember, I remember,
When Lumby had a fire,
On Shield's high verandah
Cream cans popped even higher.
But when this was straightened out
With NOCA's helping hand,
Somehow the farmer muddled through
And stayed upon his land.

I remember, I remember,
'Round Hallowe'en, one day,
A Lumby grocery man who liked
To join in pranks and play,
He took the empty cans and climbed

A telephone pole nearby
And draped the cans on many nails
From low right up to high.
T'was like a fruiting marrow vine
A funny thing to see,
But who took all those cream cans down
Is lost in history.

I remember, I remember,
The British Dairy Queen
Who toured our valley here and said
This was the best she'd seen,
And she should know, young tho' she was,
Her training had been keen.

I remember, I remember,
At Salmon Arm, one year,
The girl who's now our much loved Queen,
Walked to a thunderous cheer,
We gave her a great big NOCA cheese
To give them "Palace Cheer."*

I remember, I remember,
How NOCA did expand,
And then they called it SODICA
And spread out o'er the land
And the day it paid its millionth brick
And didn't that feel grand?

Yes, we remember the older days
When things were not so hot,
So you young farmers of today
Make much of YOUR lot!
Back SODICA for all it's worth
Give ALL THE HELP you've got.

*The Princess (later Queen Elizabeth) and Prince Philip were so interested they overstayed their schedule. The train began to move and the aides jumped on the train without picking up Prince Philip's coat and NOCA Cheese. A veteran called out "your cheese, Sir," and Prince Philip jumped off the back of the train, grabbed the cheese and his coat, waved them at the crowd and popped back on the train just as it was ready to pull out.

Milk cooperatives developed in British Columbia around the turn of the twentieth century to allow producers to share the expenses and labour of producing milk. They also allowed farmers in some areas to gain protection from the distributor-controlled fluid milk market. According to W.J. Park of Pitt Meadows, "these Western farmers had to make up their minds whether they were going to accept the lower price or let other milk come into competition. The distributor had a club over the producers."[1]

Early farmers in British Columbia were limited by the difficulties of transporting a fresh product like milk over the great distances between towns. Most farmers sold their milk very close to home. Cooperatives not only made marketing on a larger scale possible for many farmers, but they also made it possible to meet increasing calls by consumers for quality control of milk and for some measure of standardization. Returns from cooperatives provided small farmers with the financial security they needed to upgrade their herds and to expand their operations.

In 1901, there were six cooperative creameries in British Columbia: at Delta, Chilliwack, Victoria, Cowichan, Comox, and Armstrong. The oldest cooperative creamery was the Cowichan Creamery Association, founded in 1885. The largest cooperatives in the province were the Fraser Valley Milk Producers' Association (FVMPA), the Shuswap Okanagan Dairy Industy Cooperative Association (SODICA), and the Island Farms Cooperative Association. The FVMPA began in 1913 but did not do business until 1917, eventually taking in the Chilliwack, Armstrong, and Comox

Armstrong Creamery, "the home of Armstrong Cheese," 1940s.

The Creston Valley

Dairy farming in the Kootenays immediately after the war had changed little since the thirties. I arrived in Creston in the summer of 1948 following six weeks of herd management responsibilities at the Abbotsford airport after the great Fraser Valley flood. Dairy farming was one of the principle sources of farm income in the region, with income mainly derived from cream shipments. Fluid milk producers with certified premises were the minority members of the Creston Valley Cooperative Creamery Association. The industry had many inherent problems to overcome before significant economic progress could be made, including on-farm genetic deficiencies in dairy stock, nutritional inadequacies, antiquated milking systems, and above all, major marketing problems. On-farm problems could be resolved providing there was a commensurate price response from the marketing agency to offset the expense of new investments. The Co-op was only able to pay a very low price for fluid and butterfat, certainly one of the lowest prices in the province at the time. As I recall, the farmers were paid about three dollars per hundredweight, less membership deductions of several kinds. There were constant complaints about net backs of $2.75 for standard milk and, common to many areas, skepticism about the integrity of the butterfat test.

Cream shippers experienced a tremendous increase in deliveries during the early spring and summer months which tailed off rapidly at the beginning of winter. Cows were bred to freshen in late March and early April and to complete their lactations seven or eight months later. There were some days in the middle of winter that the Co-op scarcely had enough milk to meet its fluid requirement for home delivery. The creamery was a butter producing facility and most cream shippers had off-farm employment in nearby forestry or mining operations and were milking a few cows as a secondary source of family income. It was obvious that producer prices would have to increase before the imbalance in monthly milk production could be levelled out and on farm investments could be encouraged.

The creamery's marketing problem was further constrained by the start-up of a second raw milk distributor. The outlook was stagnation for the company and for the producers. I was surprised when a milk distributor from Kimberley walked into my office to talk about obtaining milk from Creston. He said, "I'm desperately short of milk every month and I could use a thousand pounds every day." That is not much milk in the context of a modern dairy farm, but it was a significant amount to the small local dairy farms in the late 1940s. I said "Before I start talking to the local farmers, I need to know what kind of money are you offering for a can of standard 3.5 milk?" He said "I will pay $5.50 delivered to the rail siding in Cranbrook." The rail costs, as I remember, were approximately fifty cents, which meant the farmers would realize two dollars per hundredweight more for their milk. I talked it over with several of the fluid shippers and was encouraged to call a meeting with the buyer in attendance. It was a rather awkward meeting as four or five of the larger producers immediately signed on to ship to Cranbrook, leaving the less certain ones with the creamery. Those shipping to Cranbrook were happy with their new and larger pay checks and talk about expanding their herds became a subject of much interest.

Sig Peterson, 2000.

creameries. SODICA began in 1925, grew out of the North Okanagan Creamery Association (NOCA), and was sold to the FVMPA in 1982. Island Farms, incorporated as a cooperative in 1944, remains the main processor of milk produced on Vancouver Island. Other regional cooperatives have existed for years and have served regional markets, including the Alberni Creamery Association, the Bulkley Valley Cooperative Creamery Association, the Cariboo Farmers' Cooperative Association, the Saltspring Island Creamery Association, the Columbia Valley Cooperative Creamery Association, the Nanaimo Creamery Association, the Nechako Valley Cooperative Creamery Association, the Kootenay Valley Cooperative Milk Products Association, the Salmon Arm Cooperative Creamery Association, the Lake Windermere Cooperative Creamery Association, and the White Valley Cooperative Creamery Association.

Other groups in the Fraser Valley, encompassing what were known as independent producers — that is, producers who had no affiliation to the FVMPA — worked together over the years to lobby for their interests and to ensure adequate markets for their milk, mostly sold on the fluid market. Interests and aims of these groups were often at odds with those of the FVMPA. Included in this group are the Twin City Cooperative Milk Producers' Association, The Independent Milk Producers' Cooperative Association, the Lower Mainland Milk Producers' Cooperative Association, The Richmond and Marpole Farmers' Cooperative Association, the Milk Shippers' Agency, and the Jersey Breeders Cooperative Association.

1906

White Valley Creamery Association
RULES

1. The Association shall collect the cream and shall pay each shareholder, out of the proceeds of the sales of butter, at a rate per pound of butter fat furnished by him or her, and shall pay non-shareholders at a less rate per pound of butter fat, both rates to be announced by notice by the Directors.

2. Patrons who may be dissatisfied with the measurements of their cream must report complaints to the Directors, who shall adjust and settle the matter.

3. The cream of each patron and of non-shareholders shall be tested; and the butter fat shall be valued according to the quantity revealed by such test and paid for out of the proceeds of sales of butter. The results of the tests shall be entered in a book furnished by the Association to the patron, and which shall be the property of the patron.

4. Each patron shall be entitled to the butter required for use on his or her own table at the wholesale price, but no quantity shall be put up in less than five pounds. Non-shareholders shall be entitled to the same quantity but at retail price.

5. In the case of a patron who does not continue to furnish the cream from his or her herd to the Creamery until the close of the manufacturing season, a sum equal to two cents per pound of all the butter fat furnished during the season shall be deducted from his or her share of the receipts, unless he or she shall have first obtained the consent of the Directors to such discontinuance.

6. The Association may insure the butter in one or more insurance companies to any extent; but the Association will not be responsible for loss on any of the butter which may be destroyed by fire, other than for the amount received by the Association from the insurance companies.

7. The cream shall be furnished from the milk of only healthy cows which have been fed upon wholesome feed with access to plenty of pure water and salt, they shall be prevented from eating any feed which will give an injurious flavour or taint to the butter.

8. The pastures, yards, and lanes shall be kept free from carrion.

9. The cream furnished by each patron shall be clean, pure and sweet, and in case any grounds should exist for suspecting that the bulk of the cream as furnished by any patron is not in every sense similar to the sample taken for use in the test, a committee appointed by the Directors shall visit the ranch or farm of the patron and make examination for themselves regarding such matter, and if any unfair or dishonest practice shall be proven to have existed, it shall be optional with the Directors as to whether they shall (1) prosecute the patron according to law, (2) effect a settlement with him or her upon the payment to the funds of the Association of such a sum as may be agreed upon, or (3) exclude the patron from the privileges of the Creamery for a stated number of years.

10. Milk must be drawn from the cows in a cleanly manner; the sides and flanks and udders should be brushed or washed, and milking with dry hands is recommended.

11. Immediately after the milk is drawn from the cow, it should be strained through a wire or cloth strainer.

12. All pails and other utensils with which the milk is brought in contact must be of tin.

13. The milk and cream must be kept in a place where the atmosphere is free from foul and injurious smells.

14. Vessels in which the milk is set must be kept clean and sweet, and the tank into which the cream cans are set shall be kept free from bad odours. If a cream collector shall discover the setting vessels or water tank of any patron to be in a state unfit for the keeping of milk or cream without detriment to its quality, he shall notify the butter-maker of that fact, who shall report the same to the patron and the Directors. After the first offence, the patron may be subjected to a fine of fifty cents for every time that a setting vessel or tank shall be found in an unclean condition.

15. Buttermilk at the Creamery shall be disposed of during the season of 1906 as the Directors determine. The cream collector, under the instructions of the butter-maker, shall reject any cream which he considers to be unfit for use in the manufacturing of the finest quality of butter, and the butter-maker's judgment in the matter shall be final.

16. All cream to be conveyed to the Creamery shall be delivered on the side of the road upon a milk stand of convenient height, and which will afford shade from the sun and protection against rain.

17. The surroundings of the milkstands shall be kept clean and free from bad smells; and the feeding of swine within 100 feet of the milkstand is strictly forbidden.

18. The cream shall be delivered on the stand at a time to suit the convenience of the hauler, who shall not leave any stand before 7 a.m. and who shall reach the Creamery with his load not later than 9.30 a.m.

19. Each patron and non-shareholder who furnishes cream to the Creamery is thereby considered as having agreed to the foregoing rules.

Handbill stating the rules of the White Valley Creamery Association, Lumby, 1906.

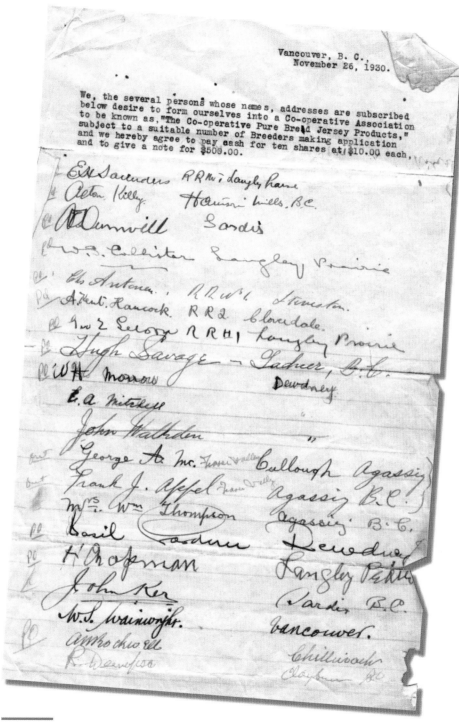

Declaration of association, Cooperative Pure Bred Jersey Products, 1930.

Cowichan Creamery

Business in Duncan and neighbouring districts continues to increase in a steady yet unobtrusive way that is apt to mislead the occasional visitor who thinks he sees in its placid existence a slumbering land of nod. The figures of the last balance sheet of the Cowichan Creamery for instance are instructive reading. During the year 1901 no less a sum than $19,651.02 have been credited to the funds, an increase of $3,403.22 on the figures of 1900. The sales of butter during the year have amounted to $80,356 pounds, an increase of 9,445 pounds, and this output has sold for $22,375.76, an increase of $3,025.46 on the receipts of 1900. In 1897 the takings were only $10,368.79, so that these have more than doubled during the past four years. The total cost to the farmer of butter manufactured by the Creamery is only three and a half cents per pound including insurance, etc. This compares favourably with Alberta figures, and the excellence of the article turned out is attested by the prizes which the company has never failed to secure wherever it has exhibited. Last year at Victoria it was awarded a diploma in addition to first and second prizes. Bravo Duncans!

The Crofton Gazette,
February 27, 1902.

Cowichan Creamery

The Cowichan Creamery Association, the first milk cooperative in British Columbia, was founded in Duncan in 1895 and was ready to open its doors for business the following year. It was financed by $10 shares from 70 shareholders. The only paid employee was the butter maker. The creamery was situated over a spring believed to contain the coldest and freshest water in Duncan — a necessity for washing the butter. In 1896, the cooperative began making butter and in the next few years expanded into other endeavours — egg hauling was added in 1909 and a few years later a feed warehouse and grain elevator were built.

A Victoria branch office was opened in 1921 and remained open until 1939. The Cowichan Creamery handled both retail and wholesale milk, but got out of the retail business by 1948. In 1950, butter production was halted. In 1988, the Cowichan Creamery Association was dissolved and its operation permanently closed.

Memories of a Cowichan Landmark

A passage from the Cowichan Leader by J.C. Harris, an original member of the Cowichan Creamery Association, 1946.

Passing and re-passing the prosaic buildings of the Creamery block, recently razed by fire, how many persons connected them with the great struggles and dramatic part they played in the development of BC? Yet they represented one of the lasting contributions to modern Canada which sprang from the real struggles of farmers and settlers just over 50 years ago. Real farming and cattle raising in the Cowichan Valley began

with the coming of the creamery. Away over in Guelph, Ont., the Ontario government had established the Ontario Agricultural College to improve farming knowledge. It came in for much ridicule — students were referred to as meddle ducks etc. — but it was destined to be of great service to Canada. Among its first professors was Professor R. W. Robertson, a fine young Scotsman, who was elevated to the new position of Dairy Commissioner for Canada by the Dominion Government.

Under the inspiration of Professor Robertson, who made his first visit to BC in 1889, the creamery question became a matter of first importance throughout Cowichan. He was a splendid fellow — tactful, friendly, a man of vision — and he grasped the needs of little farmers struggling in the backwoods settlements. He saw the difficulty of marketing the butter and eggs of the struggling settlers and with fine courage determined to devote his life to their assistance.

Homemade butter in those days might or might not be of good quality when it was presented to the storekeepers. It might be fairly fresh and well wrapped, but too often it was badly made and even going rancid, and trouble for the middlemen who had to get rid of it. What fierce rows between the farmers and the local storekeepers!

Robertson tackled his big problem right in Ottawa. There was urgent need for simple legal machinery by which the farmers could create their own cooperative associations. Cowichan Creamery was among the first to benefit from this, and from financial assistance provided. Butter and cheese were made on the farms and,

COWICHAN CREAMERY AND FARM LANDS

Cowichan Creamery and farm lands, undated.

Fall Fair on Vancouver Island

You just can't exaggerate how much the Fall Fair meant to us.

Robert Evans, 1956.

before the Creamery, each farmer sought out and established his own market — a logging camp, a hotel, or private customers. As the herd of cows grew, this primitive system became more and more inadequate. A number of creameries and cheese factories had been established in the East by the time Professor Robertson made his first visit to BC. Many graduates of Ontario Agricultural College had also come to the coast. The Provincial Agricultural Department supported the idea of a creamery here, and Professor Robertson held many meetings, supplemented by numerous informal talks and visits. The district at that time had more cattle than anywhere else in BC, except along the Fraser Valley, and it was evident that the locality was among the best prepared to start a creamery.

Farmer's wives were interested in the idea. Making butter was tremendously hard work. Churning in winter was slow and difficult, and in summer there were other serious troubles, mostly connected with the water supply. At first, the milk used to be set in shallow pans, and there were an awful number of pans to air! The deep setting system was invented; milk was lowered in a deep can into quite cold water, and cream rose much more rapidly than in the shallow pans. The skim milk was also very sweet. But, oh! The work of handling those deep cans, skimming, washing and airing. A modern trade unionist would have a fit if he had been shown all the work that the average dairy farmer's wife did as a matter of course. One of the great difficulties experienced at first was the fact that even if all the dairy farmers could be united, there were scarcely sufficient herds to support a creamery. Some already had

good customers, and were getting good prices for their dairy butter.

The actual location of the creamery was a delicate question. Cowichan was then, even more than today, a long, narrow settlement, or a series of little settlements, which, for creamery purposes, extended along the one highway from Chemainus in the north to South Cowichan. Each end began to pull hard to get the creamery located in its neighbourhood, and the controversy was hot!

It will be difficult for youngsters today to realise the difficulty of transportation 50 years ago, and how tremendously important it seemed to us to be as close to the creamery as possible. To begin with no farmers at that time had separators and it was proposed that whole milk be hauled by horses to the creamery, where the cream would be separated and the skim milk hauled back to the farms. The roads at times were almost, or even quite impassable, and it took pretty good teams and reliable teamsters and wagons to keep up those daily deliveries.

The cow population was of course the deciding factor in locating the creamery. The Chemainus end could claim Harry Bonsall's fine herd of Holsteins, first class producers. Then there were the herds belonging to the old Hubbard farm at the mouth of the Chemainus River, later Swallowfield farm. More important, at Westholme, Pat Johnston and Capt. Barkley were developing very fine clearings and many mixed cattle. At Somenos, the wonderful Horace Davis ran his fine little farm with a strong sideline in blacksmithing, and the manufacture of hay presses. Somenos was much

Share certificate, Cowichan Creamery Association, 1920.

Cowichan Agricultural Society
FALL FAIR
and Vancouver Island Sheep Fair

AT DUNCAN, B.C.

Friday and Saturday **SEPT. 18 and 19, 1936**

FRIDAY

10.00 a.m.—Fair Opens.
1.30 p.m.—Cattle and Sheep Judging.
2.00 p.m.—Children's Sports and Model Airplane Contest.
7.30 p.m.—Folk Dancing and Choir Singing by School Children.
10.00 p.m.—Fair Closes.

SATURDAY

10.00 a.m.—Fair Opens.
10.30 a.m.—Judging of Horses.
1.00 p.m.—Highland Dancing and Quadrilles.
2.00 p.m.—Judging of Light Horses, Riding and Jumping Classes.
2.30 p.m.—Loggers' Sports and Tug-of-War.
8.30 p.m.—Annual Fall Fair Dance at K. of P. Hall, Duncan.
9.30 p.m.—Fair Closes. Removal of Exhibits Commences.

Loggers' Sports and Tug-of-War
Commencing at 2.30 p.m. Saturday

Chopping Contest (open)
1st PRIZE, $15 :: 2nd PRIZE, $10 :: 3rd PRIZE, Master Mechanic Overalls, donated by Western King Mfg. Co.
A SPECIAL PRIZE of $5, donated by Jim Martin, will be awarded to the winner, provided he is wearing Master Mechanic Overalls.

Sawing Contest (open)
1st PRIZE, $12.50 :: 2nd PRIZE, $2.50 and Simonds One-Man Crosscut Saw :: 3rd PRIZE, Pair Overalls, donated by Western King Manufacturing Co.
NOTE—A Novice is one who has not won a prize in an open chopping or sawing contest. Competitors must provide their own saws and axes.

Chopping Contest (novice)
1st PRIZE, $10 :: 2nd PRIZE, $7.50 :: 3rd PRIZE, One Pair Gloves, donated by John Watson Ltd.

Sawing Contest (novice)
1st PRIZE, $10 :: 2nd PRIZE, $2.50 and Scrip valued at $5.00, donated by McLennan, McFeely & Prior Ltd.
3rd PRIZE, Pair Gloves, donated by John Watson Ltd.
A SPECIAL PRIZE of $5, donated by The Western King Mfg. Co., will be awarded to the winner, provided he is wearing Master Mechanic Overalls.

Ladies' Sawing Contest
1st PRIZE, $5 :: 2nd PRIZE, $2.50 :: 3rd PRIZE, Free Pass to Victoria or Nanaimo and return, donated by V.I. Coach Lines.

Ladies' Nail-driving Contest
1st PRIZE, $4 :: 2nd PRIZE, $2. Each competitor to drive 16 nails of various sizes. Competitors supply hammers; nails provided

TUG-OF-WAR
Teams of 8 a side. 1st PRIZE, $36; 2nd PRIZE, $18. A. A. U. of C. Rules to apply.

Highland Dancing and Quadrilles
Commencing at 1.00 p.m. Saturday. For Events and Prizes, see Prize List.

For information and Prize Lists, apply to W. Waldon, Secretary, Duncan, B. C.

COWICHAN LEADER LIMITED, DUNCAN, B.C.

Fair poster, Cowichan Agricultural Society, 1936.

favoured as the creamery site by those in the north end, whilst those in the south, favoured Duncan, or points as far south as Cornfields.

The water supply was a determining feature. There was plenty of surface water — far too much at times — but deep, secure springs were hard to locate. It was the demonstration of such springs right in Duncan, proved by the digging of wells, which made us finally unanimous in selecting Duncan as the site. The practice of hauling whole milk to the creamery, and the fact that skim milk often turned sour on the return journey, led to a scheme for keeping pigs at the creamery being tried.

Skimmed milk hardly paid for such transportation. Gradually, farmers acquired separators and the hauling of whole milk was dropped.

Nanaimo Creamery

During the middle of the nineteenth century when coal was discovered in the Nanaimo area, the population of Nanaimo swelled with mine workers and their families. For almost 20 years prior to his retirement in 1903, Samuel Robins managed the mine. An avid gardener, Mr. Robins believed that miners could gain greater stability in income if they were also part-time farmers. He made possible the subdivision and clearing of local property into 50-acre plots.

Ships coming into Nanaimo for coal added to the local demand for farm produce, so boxes and tubs of butter were readily saleable. Some miners did fulfill Robins' dream and became dairy farmers. On May 9, 1903, at a meeting held in the City Hall by the

Namaimo Farmers' Institute, it was moved by Rev. B.W. Taylor that the Nanaimo Creamery Association be formed. The motion carried and provisional directors elected for the purpose of starting a creamery were J. Randall, J. Marwick, J. Copping, T.C. Westwood, F.G. Thatcher, J. Leonard and G.L. Sahupki. William Knight of Saltspring Island was hired as butter maker in 1903 at a salary of $60 per month. In June 1906, 28 members present accepted a bid from J.C. Young to build a plant at 625 Pine Street for the sum of $980. They approved a further expenditure of $1,500 to install plant equipment.[2]

The peak output of the creamery was 200,000 pounds of butter in the 1930s. In 1946, the operations of the creamery were halted due to the decline in cream production.

The Fraser Valley Milk Producers' Association (FVMPA)

The Fraser Valley Milk Producers' Association was officially formed in 1913 when members of existing cooperatives — John Oliver and W.J. Park of Lower Mainland Milk and Cream Shippers Association; J.W. Berry of the Richmond Dairymen's Association; E.D. Barrow of the Chilliwack Creamery; and C.E. Eckert of the Chilliwack Producers' Exchange — obtained a

E.D. Barrow, first FVMPA president, 1917–1918.

DAIRYWORLD COLLECTION

Route #2

Cariboo Farmers' Co-operative Association
DAILY ROUTE REPORT
Sept 22/52

276 + 3 5
sold 242 26
Bal. 34 27.

TICKETS #51.37

	Milk		Chocolate Milk	Cream	
	Quarts	Pints		Whipping	Table
1 11 x 12	132	24		8	3
2 12 x 12	144	10	8		6
3			9		
Total Out	276	34		8	
Returned	103	4	17	8	9
Daily Sales	173	30	13	7	6
Tickets	110 + 25	24.75 2.94	4 Previous ticket coll.	1	3
Cash		7.28	Previous cash coll.	1.81	
Credit		7.39	Total Collections	4.51	
Total		37.61	Value of load 39.42	6.32	
Wholesale	3942	Qt.	Over _____ Short 1.8		
		Pt.	W.C.	T.C.	½ Pt Choc.
Fluids	12				Choc. qt
Baker bread	24		2	2	2
Floyds	36			2	3
Bettys	12	1	2	2	2
Trueman	12			2	6
	96	1	2	6	13

F W Clegg.

Route report, F.W. Clegg, Cariboo Farmers' Cooperative Association, 1952.

QUESNEL AND DISTRICT MUSEUM AND ARCHIVES

Nanaimo Creamery, undated. NANAIMO CREAMERY ASSOCIATION FONDS, NANAIMO COMMUNITY ARCHIVES

Retired FVMPA executives have a chat. Left to right: J.J. (Jack) Brown, T.M. Edwards, W.J. Park, W.L. Macken, 1960s.

GORDON PARK COLLECTION

Cariboo Cooperative Creamery, 1925.   QUESNEL AND DISTRICT MUSEUM AND ARCHIVES P1996.32.3

provincial charter for a valley-wide association. The founding idea of this amalgamation of several existing cooperatives was that "the farmer could best safeguard his ability to produce as an individual if he marketed cooperatively."[3] Before the coming of the BC Electric Railway in 1910, Fraser Valley farmers had marketed their milk in local pockets: Delta and Ladner farms were close to the cities of New Westminster and Vancouver and could serve these markets by horsepower. Upper Fraser Valley farmers sometimes sent milk to the cities via the Fraser River or the Canadian Pacific Railway, but these arrangements were difficult. The BC Electric extended the Vancouver milkshed to the eastern terminus of the railway at Chilliwack. Urban distributors capital-ized on the fact that their milk supply could come from as far away as Chilliwack in a few hours. Prices for milk fell from two dollars per can to one dollar with up to thirty cents levied per can for freight. According to William Wardrop of Deroche, "The dairies would shut off a milk shipper on a day's notice. If you went in to enquire about a market for your milk, if they did not like you, they would just say, 'We don't want your milk.' And before you were out of the door, they would pick up the phone and tell the other dairies not to take your milk."[4]

A financial depression in the province in 1913, followed by the outbreak of war in 1914, forced the organization to put off its business plans for a few years. In 1917, the FVMPA went into business with 848

PRES. JOHN HOLT, J.P.   SEC. W. COULDWELL

Cariboo Farmers' Co-Operative Association

ESTABLISHED 1921

MAKERS OF THE FAMOUS CARIBOO BUTTER

QUESNEL B.C.,   **Feb. 3rd, 1923.**

Fellow Members:-

 In presenting to you Directors report for the
operation of the Creamery for 1922 it can be truly said we have
passed through one of the worst years of trade depression and
shortage of money that this town has had for a great number of
years. The Association has feld the effect in collecting accounts. Financing has been difficult, but thanks to the Royal
Bank's assistance we have been able to get along. The farmers
have had a very difficult year, the dry season following the
damage to crops last winter affected the hay crop and pasture,
and in addition the local market for produce has been very quiet.
Had these adverse conditions not existed you Directors would
have made another call on Share Capital Account.

 During the year we have had water laid into
the Creamery from the P.G.E tank. Our sewerage system has caused
us some little trouble and we have had plans prepared and approved
by the P.G.E. Engineer to lay a drain from the Creamery to the
Quesnel River, and when conditions improve this work will have
to be undertaken. We have had a platform built in order to
facilitate loading and unloading cream from the train. It has
been found necessary to build a new ice house adjoining the
Creamery for the purpose of having ice available for cooling
cream and butter during the hot weather. The water from the
P.G.E. tank is not cold enough for the purpose. This supply
of ice will be a great help to the Buttermaker. It was the
intention of the Directors to operate the Creamery during the
winter but, owing to our inability to secure a Buttermaker
immediately after our late Buttermaker left, it was decided
to close for the winter months.

 In November a deputation from the Board of Trade
met you Directors to discuss various matters with them, chiefly
how the Board of Trade could assist the Creamery Association
in establishing new markets for butter. During the past season
we have made considerably more butter than the local market
could consume. We were unable to store it as our refrigerator
would not hold it, so we had to ship to Vancouver as the only
available market. The result has not been satisfactory from
the point of price, and it was with the idea of endeavouring

PRES. JOHN HOLT, J.P.   SEC. W. COULDWELL

Cariboo Farmers' Co-Operative Association

ESTABLISHED 1921

MAKERS OF THE FAMOUS CARIBOO BUTTER

QUESNEL B.C.,

(2)

to extend the market south between Quesnel and Clinton that
we asked the assistance of the local branch of the Board of
Trade. We believe if we could get the butter trade between
Quesnel and Clinton it would be sufficient to take care of
all our output. An effort will be made this year to get
the trade. We have already made some progress. We have customers in Pavilion, Horsefly and Chilcoten and if we can put
our price to compare with Vancouver prices I see no reason why
we cannot get all the business.

 We have been greatly assisted this year by the
Cream Special which the P.G.E. put on between Williams Lake
and Quesnel on Wednesdays. This has been of great help and you
Directors have been in touch with the Minister of Railways to
have some arrangement for this season. We have been receiving
Cream from points as far South as Lone Butte. We have had
sixty-six shippers during the year. Fifteen new members have
been admitted during the year bringing the membership to
forty-five who hold fifty-nine shares.

 We have paid to farmers for cream $10,444.30.
We made 32,261 pounds of Butter, an increase of 10,761 pounds,
an average of 3910 pounds per month during the time we operated.
Of this amount we shipped to Vancouver 13,104 pounds.

April............	1071 lbs.	Butter
May.............	2824	"
June............	5077	"
July............	5854	"
August..........	6517	"
September......	4366	"
October........	3700	"
November......	2026	"
December.......	726	"
Total........	32261	"

(3).

The Statement of Account as taken from Ledger
is as follows:-

Directors Report, Cariboo Farmers' Cooperative Association, 1923.

W.L. Macken, FVMPA president 1935–1947.

W.L. Macken

The present situation is the most critical in our history ... Conserve your present incomes because it may be a long time before we get this thing settled. The farmer asks for the cost of producing his milk and a little extra. Is there anything wrong with that? Isn't he entitled to it?... It is only by statute we are going to get anywhere, and it is only by the consent of the people who are affected that we are going to get it to work ... I'm going to battle through until we get a reasonable share of the fluid market.

FVMPA Annual Picnic, Cultus Lake, July 8, 1939.

"Aboard and Safe!"

Our member cartoonist, Norman Richardson, of Chilliwack, here shows the co-operative boat of our Association being safely navigated by our seven directors through the rocks that have a tendency to wreck it.

The Independent boat of Selfishness has evidently come to grief on the rock of Ignorance, and its only survivor looks as if he is about to perish unless he secures the lifebelt thrown by Vice-President Miller.

FVMPA members are strongly urged to think cooperatively, 1927. This is only one of many examples of political cartoons used in the milk wars between cooperative members and Independents in the Fraser Valley. BUTTER-FAT DECEMBER 1927

members organized into 16 locals: Agassiz, Dewdney, Silverdale, Pitt Meadows, Lulu Island, Ladner, Cloverdale, Langley, Dennison (Bradner and Mount Lehman), Matsqui, East Chilliwack, Camp Slough, Rosedale, Huntingdon, South Sumas, and Sardis. In these early years, the FVMPA contracted with condensing plants at Ladner and south Sumas, as well as with Valley Dairy, Standard Dairy, and Turner's Dairy. Within two years, its members had doubled their returns as they enjoyed the earnings pooled from high returns on the fluid market and lower returns from

manufactured milk. A head office was opened in Vancouver, and operations under the FVMPA banner began in plants owned by the Richmond Dairy Company, a distribution depot on Hornby Street in downtown Vancouver, at Edenbank Creamery Association in Sardis, and at the Chilliwack Creamery. Both sites were first leased and then bought by the FVMPA. In 1919, an aggressive campaign to move into the retail milk trade meant that money was needed. Members were assessed $100 per can of milk shipped daily and were required to sign notes of security. With these notes equalling

Standard Milk Company plant, 8th Avenue at Yukon, Vancouver, 1920. This would eventually become the Dairyland plant.

FVMPA

Alec Mercer, who was manager of the FVMPA, and W.J. Park in Pitt Meadows, were stalwarts in the formation of the Fraser Valley. What it started out to do was to sell milk to all the dairies and be the main supplier. The only time they all ever cooperated, in my opinion, was in 1931 when they formed the Associated Dairies. That went through and nobody heard a thing about it. It was just the most secretive thing that you could possibly think of. Not a word came out, and all of a sudden — I was employed by them at the time — I found out I had a new boss.

**Alec Mercer,
FVMPA general manager,
1933–1961.**

I can remember some of the old timers. I was never a member of the association, but I remember being invited to the annual meeting one time and I went out there. It was interesting to hear all the talk going on and then several of the other fellows who were there told me of some pretty hot meetings during the depression years. They had some strong people. Mr. Macken was a lumberman and he got brought into the association because he was interested. He was never a farmer, but he could talk their language. He was a beautiful speaker and he had an Irishman's view on everything. He could talk and control the meeting. I remember they were talking and all hell broke loose at one time. There was talk that overhead was too high in the association and he said, "Mr. Chairman, I wonder if I may ask you a question? I've heard all this talk about overhead. How high is it, and how high should it be?" They passed on to the next item just like that because they were talking about something none of them really knew.

So it was a very interesting thing. But as I say though, there are tremendous people involved. The FVMPA assisted the farmers to get a market for their milk. My father shipped milk years ago, and the only way you realized if you had a market or not is if the cans came back. I can recall one day that he took the train into Vancouver from his farm in Hazelmere. He was there, and his cans didn't come back, so he couldn't figure out what was wrong. He had that day's supply of milk, and then he had the next day's supply of milk, then he hadn't a market. The FVMPA sparked a lot of new ideas for the farmers and, to a great extent, it fought for them on the legislative level and it provided a voice that was pretty well heard over quite a large area. So, politically, it became a very viable operation for the farmers.

The Fraser Valley seemed to foot the bill for the whole thing for all those years and it seems to me any history should turn around and acknowledge that gift of the Fraser Valley Milk Producers' Association. It was quite an interesting time being in the organization.

Bill Ramsell, 2000.

$150,000, the FVMPA was able to borrow enough money to buy some Vancouver dairies — Standard Milk Dairy Company, Turner's, Hillcrest, Mainland, and South Vancouver. Bringing the retail side of the business under the umbrella of a new organization, called Fraser Valley Dairies, pitted FVMPA members against independent farmers and dealers for the business of consumers.

**J.J. Brown,
FVMPA president,
1960–1963.**

The "one man, one vote" policy of the organization meant that farm size, productivity, and the ability of all farmers to contribute effectively to the cooperative were a constant feature of debates. Some farmers favoured restrictions on membership in order to support large, efficient farms. Others favoured an expansion of membership, and the maintenance of an equal footing for all farmers, regardless of herd size, milk production, or the seasonal nature of that production. In 1923, both J.W. Berry, FVMPA president, and E.G. Sherwood, general manager, resigned and the cooperative moved in a new direction.

Under the organization of W.J. Park, of Pitt Meadows, the goals of the FVMPA were clarified: "to

**D.R. Nicholson,
FVMPA president,
1948–1959.**

Truck fleet outside Dairyland's 8th Avenue plant, Vancouver, 1938.

**J.W. Berry,
FVMPA president,
1919–1922.**

J.W. Berry

The prospects for the coming year [1920-21] are for cheaper milk. The extent to which we may be affected by depressed markets is difficult to forecast. Much will depend upon the degree of support given the association by all the milk producers of the Fraser Valley. The manner in which we have been able to cope with the very serious situation during the past few months is evidence of the value of cooperative effort. Much greater stability was shown in the markets here than in sections where no such organizations as ours operated. Those who have been producing and marketing milk in this locality in years past can quite appreciate what their position might have been with milk selling in the state of Washington at $1.70 per hundredweight for 3.8 per cent butterfat.

Annual Statement, 1920.

maintain open membership and to handle as high a volume of milk as possible; their ideal, to absorb all the milk in the valley."[5] A recruitment campaign brought membership to over 2,100. *Butter-Fat*, a publication designed to reach the farm community, kept members informed. In 1920, an evaporated milk plant was built at Delair and leased to the Pacific Milk Company. In 1924, Pacific Milk was bought by the co-op. The same year, a utility plant at Sardis was built — incorporating the equipment from the Chilliwack Creamery — to make butter, cheese, and skim milk powder.

While the rivalry with independent shippers intensified, specifically with the organization of some independent producers under the Twin Cities Cooperative Milk Producers' Association, this was an era of comparative stability and of great expansion for FVMPA members. Association members felt that it was their hard work and capital that had made such stability possible. They felt they had earned the right to good returns on the fluid market as well as the right to dispose of their surplus milk at utility plants owned by the organization. Independents, on the other hand, felt that they paid a premium just to keep milking all winter, and deserved the higher prices in the fluid milk market. Independents "resented the suggestion that they should receive the same returns as shippers who milked grade cattle for only a few months of the year."[6] In the face of these difficult issues, the FVMPA supported the passage of legislation to equalize returns to its members.

The *Produce Marketing Act* was passed in 1927. It met with strong opposition. The following year, a royal commission was established to look into the dairy in-

FVMPA members urged to buy only cooperative products, 1932. BUTTER-FAT SEPTEMBER 1932

dustry. J.W. Berry, the MLA from Langley introduced the *Dairy Sales Adjustment Act* as a private member's bill. It was passed; milk control came into effect in 1930; and a period of even greater rivalry between independents and members of the FVMPA began.

Vancouver, early 1940s, showing Dairyland's 8th Avenue plant in lower left.

71

Early FVMPA Members

Grandfather Honeyman came from Ontario to Manitoba where he stayed for a year and settled in the West in the 1880s. They homesteaded near the Gulf and after a year moved to East Delta on the old Ladner Truck Road. My father, prior to 1907, farmed on Benson Road, which is 72nd Street now. He and his brother bought 80 acres each. His name was Alec Fisher and his brother's name was Rupert. When Jack and I were married in 1940, we leased a farm for three years and then bought the farm on Smith Road (which is now 88th Street) in 1943 and farmed there until 1959. Even after we left the farm, Jack was still with the Milk Board and he and my brother had their cows together at my brother's farm. We were shipping to FVMPA and have always shipped to them.

Ruth Honeyman, 1994.

Early Challenges of the FVMPA

An interview with W.J. Park, 1956.

In 1900 I left England for Canada. I was going to St. John, New Brunswick, but after 11 days crossing the ocean we were unable to get into St. John and were finally landed at Portland in Maine. We left Portland that night, and landed in Montreal, and got on the CPR and came through to Winnipeg. I was sent out to a farm at a place called Rosser, about 30 miles from Winnipeg. On arriving there and seeing the type of farm and everything else I just wondered in my own mind whether I had made a mistake in coming to this country because it was a broken down old place, and the old fellow was going around in rags, and I wondered if there was any opportunity for me to learn anything or not. However, I had to stay with him for one year, but during the fall of that year the old man decided that he didn't want me any more. I had hired out for $15 a month and when he came to pay me he told me he only had $50 and I could take that or go without. However, I took the $50 and started away into Winnipeg. After working around in the woods and different things I came to British Columbia in 1901 in the East Kootenays and worked there for quite a while in the mines and railroad construction. I landed in Vancouver on the 1st of July. I realized then that I

W.J. Park, FVMPA president 1923-1930. FVMPA general manager 1923-1933.

DAIRYWORLD COLLECTION

THIRD ANNUAL

F. V. M. P. A. PICNIC

WEDNESDAY, JUNE 9th, 1926

—at the—

Dominion Experimental Farm, Agassiz

By kind permission of Mr. W. H. Hicks.

Mr. Hicks will be pleased to stage a stock judging contest. All buildings will be thrown open to the visitors so that the valuable stock can be seen at leisure.

The pens of birds entered in the Agassiz egg-laying contest can also be inspected.

Addresses by the Directors and other gentlemen will be given.

There will be sports for the children and a tug-of-war contest between teams from the "Locals." So all the hefty farmers want to be present. Will the locals please take this as the invitation.

The ladies' associations have kindly consented to do the catering. Lunches will be supplied for 50 cents. In case it is wet the large barn will be at our disposal.

Cheap transportation from the North side of the Agassiz-Rosedale ferry has been arranged.

If visitors will park their cars on the South side of the ferry ample accommodation will then be available.

There will be an opportunity to visit the new hotel at Harrison Hot Springs.

The grounds at the Farm are now looking a picture and all interested should make a point of visiting the newly developed health resort.

A. H. HARRIS,
Agassiz F.V.M.P.A. Local Secy.

The social side of FVMPA membership — an advertisement for the association's annual picnic, 1926.

BUTTER-FAT JUNE 1926

Staff, Comox Creamery, 1910. This plant used wood to fire the steam to run the machines until the 1950s.

From Four Cans to Two

Before the FVMPA was formed, we were shipping four cans of milk to Turner's Dairy, and old Turner came up and told Dad he was only taking two cans. Dad says, "What will I do with the rest of it?" Turner said, "Throw it in the ditch for all I care."

Glenn Toop, 2000.

had struck the country I had been looking for and anticipating.

I worked on farms for two or three years down on Sea Island and Lulu Island and then finally decided I would try and locate a farm for myself. After viewing farms in Sea Island, Lulu Island and Ladner, I finally drifted to Pitt Meadows in 1905. I purchased the farm with my father-in-law, Mr. John McMyn; we bought some 300 acres of land here and started in a partnership. After one year we dissolved the partnership and divided the farm and I started in on my own, and have been carrying on ever since. Of course when I came here, this country was in its primeval state. There was particularly no farming being done here at all, except on some of the higher ground. As far as the low ground was concerned there was no farming at all, and all the high land I had to clear to make it into a farm. Also the low land was covered with scrub willow and crab apple and that all had to be cleared and after it was cleared, drainage had to start in.

After getting a little of the land cleared I purchased three cows at $35 a piece and started shipping half a can of milk with others to the dairy in Vancouver. I gradually increased my herd till I was milking about 12 cows and shipping about two cans of milk to the dairy in Vancouver. The barn that I had to stable these cows in was practically built from the bush. All the poles, rafters, shingles, and the siding were split cedar. The inside of the barn was split cedar for planking — a barn that today I'm afraid wouldn't qualify for shipping to the Vancouver market. We had quite a job in trying to grow tame grasses on the low land owing to

the land being very wet, and we were dependent for quite a while on the wild grasses that grew. However, after some four or five years, we gradually began to get the tame grasses to come in. I might say that the wild grass that was cut here was not of very good quality. It would take two or three tons per animal to keep them moving at all, and it was necessary to buy concentrates because up to this time we had been unable to grow oats of any value. Mostly we grew the oats and cut them green and made them into hay, as the oats were too light if you permitted them to go to harvest. However, with the years, and with drainage and cultivation, we can now grow crops comparable to anywhere in the Valley.

I started shipping milk to the Vancouver market and the dealer established the price. The price at that time was about $1.50 a can, and when the summer came along we got a letter from the dealer telling us to stop shipping milk or just ship cream, and probably we would have nicely got to the point of shipping cream and got a few hogs to feed the skim milk to, when we got another letter telling us to ship milk. The conditions at that time were very, very bad as far as the dairymen were concerned. An organization was started which was known as the Lower Mainland Milk and Cream Shippers Association. This was just a bargaining association, and Mr. Buckingham, the father of Mr. Buckingham of Rosedale, was the secretary of this organization. I joined it some time later and was appointed on a committee to meet the dealers in the City of Vancouver, to discuss the question of price with them for the coming winter. We met on two different occasions

74

Comox Creamery, circa 1910. This plant was built in 1901 for $4,040.

FVMPA Board of Directors, 1921. Front row, left to right: J.W. Berry, W.J. Park, Alex Davie. Back row, left to right: Lloyd T. Beharrell, A.H. Mercer, J.W. Miller, J.F. McCutcheon.
DAIRYWORLD COLLECTION

and discussed the price for the winter and then reported back to our organizations. The price, if my memory is correct, was 50 cents per pound butterfat, but unfortunately when the first cheques came in for our winter shipments, the price was only 40 cents instead of 50 cents. This aroused quite a lot of feeling among the dairymen and I might say at this time, that all the milk available in Vancouver was produced on Sea Island or Lulu Island and on the north side of the Fraser, from about Agassiz down. There was nothing coming in to amount to anything from the south side of the Fraser. However, the organization finally began to realize that as a bargaining association, they were not going to get very far, and in 1910 there was a general meeting called in Westminster and the Fraser Valley Milk Producers' Association was first mooted. The men who met at that time were Mr. John Oliver, afterwards premier of the province, Mr. E. D. Barrow, who became Minister of Agriculture, Mr. J.W. Berry of Langley, Mr. C.E. Eckhart of Chilliwack, Mr. Vanderhoof of Huntington, and myself.

At that time the government had brought in the Cooperative Act *and we had studied this* Cooperative Act *and decided that probably we could come under the act and form an organization that would handle our milk direct from the producer, through our organization to the dealer, without the individual farmer having any contact with the dealer whatever. However, we started out with the idea that we would be able to put this over, and travelled through the lower part of the Valley, contacting the farmers, most of it by horse and buggy or horseback, or by the steamboat on the river. Just about this time the BC Electric opened up their line to Chilliwack, and opened up another avenue of milk for the fluid market in Vancouver. The dealers immediately took advantage of this new supply of milk which they could purchase at a lower price than ours down here, and they came to us and offered us a price of five or six cents above the price that they could secure from the Chilliwack Valley. I might say that very little, if any, of the Chilliwack milk came down to the Vancouver market owing to the fact that transportation wasn't as convenient as the system which was prevailing in the lower part of the Valley; however, the fact that that milk was available was utilized by the dealers to cut the price of the milk down here. In*

Comox Creamery, 1910.

A Little More
and a Little Less

A little more kindness, a little
    less creed;

A little more giving, a little less
    greed;

A little more smile, a little less
    frown;

A little less kicking a man when
    he's down;

A little more we, a little less I;

A little more laugh, a little less
    cry;

A little more flowers on the
    pathway of life;

And fewer on the graves at the
    end of the strife.

*Written for FVMPA members
by "a Matsqui Farmer," 1932.*

our efforts to form the Fraser Valley Milk Producers' Association, it was then necessary to widen our field and go into the Chilliwack Valley. We decided on holding a meeting in the Chilliwack area at a place called Cheam and we had a fairly good meeting — discussed the whole situation which was more or less foreign to them as they had always been operating just a small creamery cheese factory in the Chilliwack district. When it came to the question of the fluid market it opened up a new avenue which looked very, very good to them. We covered the Valley — of course you must remember it was slow travelling in those days — we couldn't fly from one end of the Valley to the other. The success we had hoped for amongst the farmers in trying to organize this Association was not there simply from lack of education and lack of close contact with farmers, so we delayed our efforts until 1913, when we again made an attempt and went so far as to secure a charter from the provincial government under the Co-operative Act *known as the Fraser Valley Milk Producers' Association.*

We then got busy again and went out into the country to see if we could sign the farmers up and after some three or four months of effort found we couldn't get a sufficient number, which in our opinion was necessary before we could attempt anything such as the Association pertained to be. We drifted on then for the next two or three years until 1916 when the situation had become considerably worse as far as the farmer was concerned, in that prices for all of the things that he required was going up owing to war conditions, while the price of milk still stayed down. In 1916, we can-

vassed the Valley from Ladner to Hope in teams of two men and finally were successful in securing sufficient members in our opinion to start the organization.

In 1917, in the month of February, we decided to launch our Fraser Valley Milk Producers' Association. When we did start it, some of the dealers who were very much upset had gone out into the country to see their shippers. It was very lucky for us that the ones they went to see were signed up, as I feel sure that if they had gone to some of the others who hadn't been signed up we probably would never have started the Fraser Valley Milk Producers' Association, at least at that time, because the percentage that we had, was not more than about 60 percent of the producers in the Valley. However, we took the risk of starting it, and after the dealer who had been out and visited his dairymen found that they were signed up with the Fraser Valley, he accepted our organization.

There was a fellow named Murdo McLean from Dewdney; and a fellow named Hodgson from Dewdney, who operated a farm at Hatzic; and there was a Mr. Cooper from Silverdale; Jack McCutcheon, who ultimately became a director; Mr. B.A. Harrison of Langley; and Alex Davie of Ladner. There were just about a dozen of us all told including the provisional board that canvassed the country for members.

I went in there as general manager in 1923 and in about 1925 prices got desperate and dealers had gone out into the country offering any shippers who would ship to them five cents a pound butterfat more than the association could offer, owing to the fact that we were doing the manufacturing of any surplus at that time.

The new Comox Creamery plant, built in 1946.

South of Ladner

My father came out from Nova Scotia in 1910. Dad started out as a charter member of FVMPA in 1916. He farmed on Parmiter Road, two and a half miles out Boundary Bay Road, south of Ladner. He first farmed on the Skinner Farm with his brother until 1920, then he rented the farm on Parmiter Road for two years and then he bought it. We still farm it but we lease it back from the government. We were expropriated for 4,000 acres as back-up land for Roberts Bank.

We have always shipped to FVMPA. My father remembered the early days when often the milk from the Independents was dumped when there was too much available. And for that reason, he stuck with the FVMPA because he said maybe we didn't get too much for it but we didn't have to dump it. The Co-op had a powder plant, a cheese plant, and the Pacific Milk plant, so there was always a place to put the surplus milk. The price was about 90 cents per hundred pounds then.

A. Bates, 1993.

The membership of the Fraser Valley decreased very considerably, and it became alarming to the then Board of Directors, and it was decided to make a canvass of all old members and any new men who were establishing themselves in the Valley, with the idea of trying to get a 100 percent Fraser Valley organization. We travelled from one end of the Valley to the other in groups of two, and called on every individual farmer that had previously been a member, or any new farmer that we were advised of, and during that six-month period, we were successful in signing up about 90 to 95 percent of all of the dairy farmers in the Fraser Valley. We felt satisfied at that time that we had accomplished what we had set out to do, and that was to create an organization that every farmer in the Valley would ship to. However, our expectations were dashed to the ground when the Depression came along, and with prices tumbling all over and the dealers still offering an advance over what we would pay, and the financial condition the farmer found himself in, we very soon lost many of the men we had signed up in the earlier period, until we were down at one time to about 65 percent of the shippers in the Valley.

The trouble was the Depression was right on us and a dollar meant so much to anybody that you couldn't blame them. The fact was that the dealers, if my memory is right, had raised their price from five cents to seven cents a pound butterfat more than we could pay. By this time we were supplying no milk to the dealers at all. The dealer was getting his full supply from independent farmers in the country.

About this time there was a strong agitation amongst the members of the Fraser Valley Milk Producers' Association to cut the price of fluid milk to a price that would bring a return to the Independent, about the same return as the FVMPA could receive. However, although it was argued very strenuously, both by many of the members and by some of the directors, there were a few of us who were not agreeable to do this, because we could see that there never would be a time while we had a surplus that we could lower the price sufficiently low that we could meet the dealers' price. I think it was Mr. Dodsley Barrow, if my memory is correct, who argued most strenuously that we could never meet the independent distributor. We would ruin all our own farmers if we ever attempted to do it. However, for a short period of time the price of milk was cut on the fluid market and it clearly demonstrated the fact that although we were selling 12 to 14 quarts of milk to the dollar, the independent dealer could still pay more than we could for our total production of both manufactured and fluid. I am of an opinion that to try and attempt anything like putting our surplus on the market at a price that would bring the independent distributor to time would wreck not only them, but us and all. I don't think it can be done.

The consumer at that time was being fed so much stuff by the independent distributor that I think that as far as the Fraser Valley Milk Producers' Association was concerned, we had a pretty tough fight with consumers. There were scores and scores of them that would ring me up when I was in there as general manager and bawl me out for the condition of the milk and everything else. We put two ladies on and they went around

FRASER VALLEY MILK PRODUCERS' ASS'N.
703 ROGERS BUILDING
VANCOUVER, B.C.

April 1922

Jas. Erskine, Jnr.,   1870

Eburne,

B. C.

*Below is a statement showing amount due you for 1921
Deferred Payment of 2¼c. per lb. B. F., and Interest on your
payments on shares at the rate of 8% per annum:-*

1921 Deferred Payment

3133 lbs. B.F. @ 2¼c.   - $ 81.74

Interest for 1921   - $ 27.56

Total   - $ 109.30

In settlement of the above you will find enclosed:-

Certificate No. 6236 $ 40.—

Cheque No. 8963 $ 69.30

Total   - $ 109.30

*The Annual Meeting voted the payment of approximately $33,000.00
in cash, and $30,000.00 in stock.   The above calculation is based on these
instructions, bearing in mind the fact that the stock must be in multiples of ten
dollars.*

Annual statement to members, FVMPA, 1922.
TOM ERSKINE COLLECTION

BUTTER-FAT
Stands for Better Farming, Better Business and Better Living.

A FARMER-OWNED FARM JOURNAL FOR
CO-OPERATIVE DAIRYMEN

Published Monthly by

**The Fraser Valley Milk Producers' Association
Vancouver, B.C.**

P. F. COLLIN, Editor

Subscription Price to Non-members, $1.00 per Year
Mailed to Any Address

ADVERTISING RATES ON APPLICATION

Editorial information, *Butter-Fat*, August 1926.

When the Boys Came Home

"Keeping the home fires burning" today is just as important as it was in the days of 1914-18. Some of the 500 ex-service men who form part of this organization of 3,000 milk producers are doing a little different kind of digging to what they did in those war days. It's ditches instead of trenches, combating weeds instead of whizbangs, and pumping lacteal fluid — the world's greatest food and life saver, instead of doing the reverse when they were wont to pump something of a dangerous character.

It is necessary for our present and future well-being to "Keep the Home Fires of this Association Burning." Keep the Business of this Association prospering, hold this organization of 3,000 milk producers together and foster the spirit of cooperation — the spirit that won the war in bygone days.

One sure way of putting the "Home Fires" out — one certain method of destroying this business that has taken more than a decade to build up, is to fail to cooperate, to patronize the other fellow's product, and to buy dairy products which are made from Independent milk.

Butter-Fat, *June, 1928*

and visited the different homes where we had telephone calls as to the quality of milk or some other reason, and we found from these ladies' reports that our great trouble was that the Independents were putting out all the lies that they possibly could think of about the Association, and about the milk. We were able to combat a tremendous amount of the propaganda that was being advertised as far as the Fraser Valley Milk Producers' Association was concerned. But there was no question in my mind that we were in very bad repute as far as the milk situation was considered in the city of Vancouver because the propaganda turned the mind of the consumer against the FVMPA and for any rise in price that was made in the Vancouver market we were immediately blamed.

Today I think the Fraser Valley are accepted pretty nearly 100 percent by the consumer, but in those days it was tough I'll tell you, very very tough. One fellow happened to call on my wife's sister — selling milk. She said, "Oh, I'm taking Fraser Valley Milk." He said, "That stuff's rotten." That was an actual talk between them. He said, "Fraser Valley only have little barns and places out in the country where they just have a few cows." He said, "their milk is dirty" and everything else. "Well," she said, "you can't tell me that, I know different."

The Fraser Valley Milk Producers' Association decided that I should go to England. We had agents over there, we were making quite a lot of powder then and there wasn't a very big market for it so they decided that they would send me over there and at the same time the Associated Growers wanted to send Mr.

Chambers over there to look at the markets of the fruit men. It was arranged that the two of us would go together. I was given a bunch of introductions over there, and I thought they were useless, I didn't think they were any good at all. But you know that old devil Captain Dunwaters gave me those introductions and I'd never have got to see those people if it hadn't been for them. You had to have them. You couldn't barge in like you could in this country, so I took all these names. I got a hold of a firm in London and went in to see them. I told them I had some of our product with us and asked if we could sell them any. I think maybe I had sold 100,000 pounds of powder or something like that, I don't remember the amount, but anyhow what I was primarily interested in was selling this big stock of Columbia Milk, so this fellow told me to go to Birmingham and see a certain fellow there, and he gave me an introduction to him. He said he was a big operator, so I went to Birmingham to see this fellow and he sent me from Birmingham to Bristol to see another fellow there. The Bristol man recommended that I go to London and see the head man in London.

So I went to London and called on this fellow and he was rather stuck-up, at least I thought, a kind of a la-de-da sort of a fellow. I thought, well, I'm not going to get anywhere now, after travelling all over this darn place and one thing and another, so he asked me into his office. He said, "You come from British Columbia — Vancouver." He talked to me and he asked me all about Vancouver and what kind of a country it was and everything else you know, and never broached anything about the business. I thought he was just inter-

Schmidt, John P.
R.R.1, Sardis.

Schroeder, Jacob P.
R.R.1, Sardis.

Schroeder, John
R.R.1, Sardis

Schum, Adam
Ladner.

Scotney, H.
R.R.1, Coghlan.

Scott, Andrew
Murrayville.

Scott, John William
R.R.2, Cloverdale.

Scott, Peter L.
1548 Kitchener St.,Vanc'r.
Farm: R.R.1, New Westminster.

Scott, Robert E.
R.R.1, Abbotsford.

Sepass, David
Sardis.

Severinski, S.
Port Hammond.

Shaw, Harold Cordon
R.R.1, Milner.

Sheridan, Chas. Ambrose
Langley Prairie.

Shuster, Mike
R.R.1, Coghlan.

Sidaway, Annie & T.
R.R.1, Steveston.

Siegrist, Ernest
R.R.1, Aldergrove.

Siemens, David J.
Yarrow.

Simpson, Mrs. Isabella
Port Coquitlam.

~~Singh, Gokal~~
~~Pitt Meadows.~~

Singh, Gunga
Abbotsford.

Siviski, Pete
Aldergrove.

Skea, James
Coghlan.

Slivens, Mrs. Mary
R.R.1, Chilliwack.

Sluis, Mrs. Lammert,
Pitt Meadows.

Smillie, William
R.R.1, New Westminster.

Smith, Archibald
R.R.2, Kitchen Rd.,
Chilliwack.

Smith, Charles
Box B, Ladner.

Smith, Elizabeth
Matsqui.

Smith, George H.
R.R.1, Agassiz.

Smith, John Richard
Aldergrove.

Smith, M. Plaxton
c/o Miss Alice Green,
R.R.2, Langley Prairie.

Smith, Sam
R.R.3, Cloverdale.

Smith, William Hamilton
R.R.2, Langley Prairie.

Smyth, Albert
Huntingdon,

Snow, Fred C.
R.R.1, Chilliwack.

Soberg, Mrs. P.
812 Ewen Avenue
New Westminster.

Somerville, John Harvey
R.R.1, Langley Prairie.

Somerville, Richard F.
R.R.1, Port Haney.

Sowden, Johnson
Surrey Centre, P.O.

Sparrow, George
Huntingdon.

Spires, Robert H.
R.R.1, Eburne.

Sprengel O.
Sullivan Station.

Spring, Isaac T.
R.R.1, Matsqui.

Stade, Clarence Walter
Box 335, Chilliwack.

Steel, James
Abbotsford.

Steppler, Valentine
Pitt Meadows.

Stevens, Francis E.
R.R.3, Sardis.

Stevens, Hector
Abbotsford.

Stevens, Joseph
R.R.3, Cloverdale.

Stevens, P. Barnard
Langley Prairie.

Steves, William C.
No. 1 Rd., Steveston.

Stewart, William
R.R.1, Coghlan.

Stickney, William
R.R.1, Milner.

Still, Ernest John Darroll
R.R.1, Cloverdale.

Stillhamer, John
Murravyille.

Stirling, Sam
Murrayville.

Stoneman, Frederick
R.R.2, Chilliwack.

Storey, Frederick Charles
Chilliwack.

Stovell Brothers (S.M. & E.)
R.R.1, Langley Prairie.

Strohmaier, Mrs. Mary
Chilliwack.

Stromberg, Fritz
Huntingdon.

Strudwick, Ralph
Murrayville.

Sturtz, Phillip
R.R.1, Steveston.

Sumpter, William Henry
Box 115, Agassiz.

Susani, Mathew
R.R.1, Mount Lehman.

Swinden, Mrs. Ellen
Whonnock.

Sykes, Norris
Pitt Meadows.

Tanner, Jack O'Malley
Dewdney.

Taylor, Mrs. Cora
R.R.1, Abbotsford.

Taylor, George Herbert
Mount Lehman.

Taylor, James
Chilliwack; B.C.

Taylor, William Lee
Deroche.

Tecklenborg, Ludwig
Milner.

Terpenning, Martin E.
R.R.1, Langley Prairie.

Tetz, Friedrich
R.R.1, Chilliwack.

Theimer, John
R.R.1, Chilliwack.

Thiessen, Daniel E.
Yarrow.

Thiessen, Susanna (Mrs.)
Yarrow.

Thompson, James
R.R.2, Cloverdale.

Thornton, George Edward
Sardis.

Thornton, Geo. Israel
Sardis.

Thorpe, Alfred Henry
Pitt Meadows.

Threlfall, Harry
R.R.1, Agassiz.

Thrift, John C.
R.R.2, Cloverdale.

Thrower, George Walter
Mount Lehman.

Timms, John
R.R.1, Sardis.

Tingle, Robert L.
R.R.2, Sardis.

Tocher, Alexander Dow
R.R.2, Cloverdale.

Tocher, Francis S.
R.R.1, Cloverdale.

Toews, Abram D.
Rosedale.

Toop, Gordon H.
R.R.1, Sardis.

Toop, Jack H.
R.R.1, Sardis.

Towlan, James
Aldergrove.

Towle, Geo. Clayton
Milner.

Townsend, Harvey
R.R.1, Milner.

Trainor, John Chas.
R.R.1, Ladner.

Trammer, Alfred
R.R.1, Sardis.

Tribe, Kenneth Julian
R.R.2, Chilliwack.

Triggs, Arthur Chas.
R.R.1, Port Kells.

Tuyttens, Jerome
Agassiz.

Tuyttens, Raymond Geo.
Agassiz.

Twemlow, Wm. Ed. D.
Coghlan P.O.

Uchida, Mitsugu
Durieu.

Vajda, Andrew
Box 20, Pitt Meadows.

Vandale, Mrs. Anna
Agassiz.

Vanderveen, Clarence
Howes Rd., Glen Valley
Vandrishe, George
R.R.2, Cloverdale.

Vaughan, Victor Clarence
R.R.2, Langley Prairie.

Vickman & Jackson,
R.R.1, Langley Prairie.

Viellechner, Ludwig
R.R.1, River Road,
New Westminster.

Vincent, Henry
R.R.1, Abbotsford.

Vopnfjord, Jacob
R.R.2, Cloverdale.

Vrba, Joseph
R.R.1, Sardis.

Wall, Jacob J.
R.R.1, Sardis.

Wandfluh, Mrs. Bertha
R.R.3, Sardis.

Ward, Harold
R.R.1, Langley Prairie.

Warren, James
DeWolf Ave., Chilliwack.

Wasylenchuk, Mrs. Annie
R.R.3, Sardis.

Watson, Edith E.
R.R.1, Sardis.

Watson, George (3)
Box 17, Milner.

Watt, James Wilfred
Murravyille.

Waugh, John
R.R.1, Langley Prairie.

Weatherby, Clement C.
R.R.1, Aldergrove.

Webb, John
R.R.1, Ladner.

Webb, William W.
R.R.1, Milner.

Webster, James Anthony
R.R.2, Sardis.

Webster, Marion L.
Atchelitz P.O.

Weir, James
R.R.1, Milner.

Welsh, David William
R.R.1, Milner.

Welton, George,
P.O.-Murrayville.
Farm-Hopington.

Wengerchuk, N.
R.R.1, Milner.

Westlin, Mrs. Josephine
(Exec. D. Roussel Est.)
P.O. Box 162, Agassiz.

~~Weston, John~~
~~R.R.1, Chilliwack.~~

Whelpton, Ruby May
Agassiz.

Whitcomb, John Howard
R.R.1, Milner.

White, Harry L.
Matsqui.

Whiting, Richard Sidney
Whonnock.

Whitlam, Mrs. Margaret A.
R.R.2, Prest Road,
Chilliwack.

Whitson, William Ruxton
R.R.1, Matsqui.

Whyte, William Watt
R.R.2, Chilliwack.

Wiebe, John A.
R.R.1, Agassiz.

Wiens, Jacob G.
R.R.1, Sardis.

Wiens, Jacob J.
R.R.1, Sardis.

Wiens, John G.
R.R.1, Sardis.

Wiens, Wilhelm
R.R.3, Sardis.

Part of the FVMPA shippers list, 1940s. ACTON KILBY COLLECTION, KILBY FARM MUSEUM

Cream Stolen

On Saturday morning, some person or persons took a can which Mr. W. Hobbs of Salmon Arm had left in its usual place for John McPherson to pick up. It had been raining during the night. The tracks of the vehicle in which those who took the can of cream were riding were clearly visible.

No doubt with the present stringent butter ration of six ounces per week, a nice can of fresh cream looked pretty good to some butter-hungry people. It is suggested that patrons be careful and do not leave their cream sitting by the roadside too long.

The Cream Collector, *July 1946.*

ested in knowing something about the country. Finally he said, "You know, we've tried that milk of yours, and we're rather satisfied with it. How many cases have you got?" I don't remember how many cases we had, but it was something like 150,000. The old fellow said, "We'll try 100,000 cases." It was a funny thing, this happened in 1923 and in the fall of 1924 we paid the banker off. We didn't owe the bank a darned cent.

Shuswap Okanagan Dairy Industry Cooperative Association (SODICA)

The beginnings of SODICA reach back to the turn of the century to a time when a series of creameries existed in the Okanagan, each processing milk from local farm-

T. Everard Clarke, manager for over forty years of the association that became known as SODICA.

LLOYD DUGGAN COLLECTION

ers. In 1902, a creamery was built one half of a mile north of Armstrong, called the Armstrong Creamery and operated for some time under the auspices of the North Okanagan Creamery Association (NOCA). By the early 1920s, the creamery was falling on tough times; indeed, "overhead and lack of good sales cut too deeply into the profit to survive much longer."[7] In June, the members got together to discuss their

financial situation and to consider two proposals that were on the table: one from the FVMPA and one from the Pat Burns Company. The offer from the Pat Burns Company was accepted, and Burns took over control of the creamery at Armstrong on July 1, 1925. The Association was now known as the Okanagan Valley Cooperative Creamery Association, but it retained NOCA as a brand name. Everard Clarke was hired as manager, and the first Board of Directors included C.J. Patten, R.J. Coltard, R.A. Copeland, W.S. Cooke, Thomas Gray, Major P.J. Locke. In 1925, 385 shippers produced 338,301 pounds of butter. By March, 1927, three butter makers were employed by the association and 450,000 pounds of butter were manufactured. In September of 1927, the Armstrong Creamery was destroyed by fire and operations were moved temporarily to Vernon. The following year, the decision was made not to rebuild in Armstrong, but instead to plan an expansion of the Vernon plant, which was completed in 1936. In 1928, the association acquired a second plant in Enderby and in 1944 purchased Royal Dairy in Vernon. SODICA was managed for over 40 years by T.E. Clarke, a "good manager and a good leader who has done much for the farmers of the Okanagan, not only as their manager, but by introducing many progressive innovations and techniques for the welfare of the farmers and the Okanagan dairy industry."[8] He started the in-house magazine called the *Cream Collector* in 1927 under the banner, "We Are Not Milking Cows for Honor and Glory."

In May of 1947, the association began operating under the name Shuswap Okanagan Dairy Industry Coop-

Board of directors, Okanagan Valley Cooperative Creamery Association, 1930. Back Row, left to right: E. Clarke, S. Halksworth, J.R. Freeze, J. Gillian. Front row, left to right: R. Peters, R.J. Coltart, Charles Patten, J. McCallan. Some of these men served the dairy industry in the Okanagan for a long time. R.J. Coltart was president from 1936–1939. Sam Hawksworth was president for almost twenty years, from 1939–1958, and was with the association when it adopted its new name, the Shuswap Okanagan Dairy Industry Cooperative Association, in 1947.

SODICA shippers make good-will tour to Kamloops to demonstrate the importance of the dairy industry in the Okanagan, 1960s. From left to right, Okanagan historian and dairy farmer, Beryl Wamboldt; Falkland dairy farmer, Mrs. John Babij; and Mrs. Reg Saunders.

LLOYD DUGGAN COLLECTION

erative Association after an agreement was reached to amalgamate the Salmon Arm Creamery Association, which had organized in 1915, and the Okanagan Valley Creamery Association. The NOCA name was again retained as the product brand name. In 1951, the Salmon Arm plant opened and became a butter and cheese manufacturing plant. The following year, the Kelowna plant went into operation.

The decision of the Okanagan Valley Creamery Association not to rebuild its operation in Armstrong after the fire in 1927 paved the way for Armstrong area shippers to open their own creamery in Armstrong.[9] In 1938, the creamery was completed, and Charlie Busby was hired as cheese maker and manager. The following year, the group incorporated as the Armstrong Cheese Cooperative Association.

NOCA plant, Enderby, 1940s.

Left to right: Ed Stickland Jr.;
Mr. E. Emeny Sr.; Ed Stickland Sr.;
Hugh Mason, federal grader,
circa 1960.

Farm of the Year,
1954, Yew Tree
Farm, Grindrod.

Looking for Greener Pastures

Des Hazlette and Jim Ryder, District Agriculturists of Salmon Arm and Vernon respectively, did an excellent job throughout the recent SODICA Green Pasture competition. Together with Bill Cameron of SODICA, they made the arrangements for the two-day tour and pasture judging competition. This competition is open to any dairy farmer in the Okanagan. It was the consensus of opinion that this was the best tour since its inception nine years ago. The efforts of these two hardworking men, without whose interest and drive projects such as this could not be carried out successfully, are greatly appreciated in all phases of agriculture in the Okanagan.

The Cream Collector, *September 1962.*

Ed Stickland Sr. on horse, R.J. Coltart and T.E. Clarke on fence, Enderby, 1934. THE HISTORY OF SODICA

SODICA director, Ernie Skyrme of Yew Tree Farm, Grindrod. Along with his family, Mr. Skyrme was declared winner of the 1954 Farm of the Year award from the BC Power Commission.
LLOYD DUGGAN COLLECTION

Ed Stickland, president of **SODICA** for many years, mid-1960s.
LLOYD DUGGAN COLLECTION

SODICA truck fleet, Vernon, 1949. Left to right: Erwin Klingspoon, Jack Fuhr, Peter Palm, Bill Skobalski, Gordon Watson, Wally Bennett, Mac McKenzie, Larry Antilla, Bill Cameron.

George Malcolm (left) and Norman Ingledew (right), founding members of Island Farms, 1946.

FRED MOCKFORD COLLECTION

A.E. Warner and A.E. Sage were the main shareholders. The Board of Directors included A.E. Sage, Jack Evans, Edgar Docksteader, A.E. Warner, E.A. Norman, and Bert Pritchard.

Island Farms Dairies Cooperative Association

George Malcolm and Albert Doney, brothers-in-law, began Registered Jersey Dairies in the early 1930s from simple beginnings, according to Ron Greene, a historical researcher who has pieced together the history of the partnership.[10] Malcolm produced and bottled the milk from his herd of pure-bred Jerseys at the family farm on the foot of Mt. Newton in central Saanich. Doney de-

livered the milk, and together they expanded their operation quickly. At first, they called their partnership the El Sereno Dairy. As the years passed and business grew, the home farm could no longer supply enough milk for the retail route, so the partners began to buy milk from other producers and decided to rename their operation. By 1934, the Registered Jersey Farms name was in use. The following year the partners opened a depot and an office at 608 Broughton Street in Victoria and were supplying milk to 200 families on a 67 mile route that Doney serviced by leaving home at 10:00pm. In 1937, Registered Jersey Dairies Ltd. was incorporated with three shareholders, Malcolm, Doney, and their accoun-

Board of Directors, Island Farms, 1947. From left to right: Robert Cheyne, auditor; Harry Dawson, Nanoose; George W. Malcolm, secretary and general manager; W. W. Michell, Saanichton, president; Captain C.L. Anderson, Cobble Hill, vice president; John E. Martin, Sooke; A.W. Aylard, Sidney; Fred Wilson, Cedar.

Supporting Young People

Increasing interest is being taken in Calf Clubs throughout the Island. Island Farms Cooperative Association has donated $25 in cash prizes for Calf Club exhibits at the Ladysmith Fair. Youngsters under sixteen years of age, showing calves of any breed which were born this year, will be eligible to compete for these prizes.

Island Farm News,
June 1947.

tant, W.D. Osborn. Along with distributing the milk from Malcolm and Doney's herd, this company handled the milk of A.W. Aylard, Major A.D. Macdonald, Miss E. Moses, Ian Douglas, and P.J. Jeune. In the next few years, more shareholders were added; however, capital was short. In 1942, Island Farms Ltd. was formed and the shares in Registered Jersey Dairies Ltd. were transferred to the new company, along with the plant and equipment, seven delivery trucks, and the company's lunch counter. Three businessmen from Vancouver became shareholders, bringing with them the required financing: Philip Fleming, William E. Hammond, and N.H. Ingledew held the majority interest in Island Farms Ltd. with almost 80 percent of shares. Expansion plans were ambitious, but money was tight.

After less than two years under Island Farms Ltd.,[11] some of the farmer-shareholders became disgruntled because they felt they had lost control over their milk sales. They appointed a provisional Board of Directors consisting of Arthur Aylard, George Malcolm, W.W. Michell, Captain Gibson, and Andy McGregor to set up Island Farms Dairies Cooperative Association on January 1, 1944. In order to raise the $20,000 down payment required to buy Island Farms Ltd., members of the new co-op signed promissory notes agreeing to pay $200 per can per daily shipments, an arrangement carried on for the first few years of the co-op's operation. At the first annual meeting of Island Farms Dairies Cooperative Association, a new board of directors was elected: Arthur Aylard, president; George Malcolm, Secretary; W.W. Michell, vice-president; C.L. Anderson, D. Bennie, and H.D. Evans, members-at-large.

Incorporation papers, Registered Jersey Dairies, 1937.
FRED MOCKFORD COLLECTION

Island Farms

Donald McKinnon, 1993.

I worked for Island Farms from 1945 until 1951. Before the war, I worked in the accounting department of the Imperial Oil Refinery in Regina, Saskatchewan, and from there I joined the navy in May 1940. It was my intention to return to my job, which was the arrangement made by Imperial Oil — that they would hold all the positions open until the end of World War

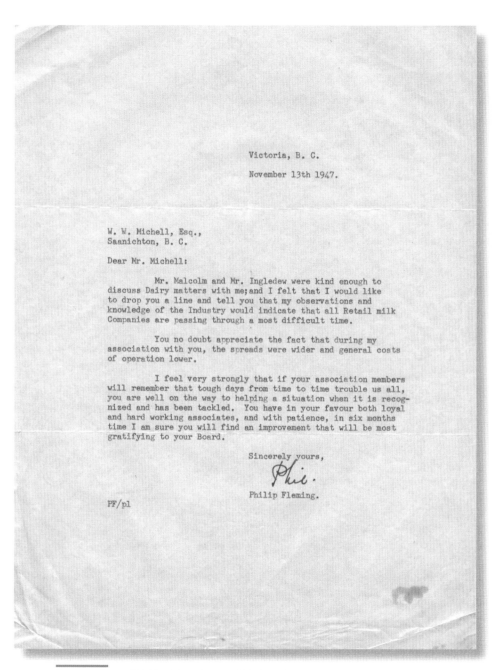

Victoria, B. C.

November 13th 1947.

W. W. Michell, Esq.,
Saanichton, B. C.

Dear Mr. Michell:

Mr. Malcolm and Mr. Ingledew were kind enough to discuss Dairy matters with me; and I felt that I would like to drop you a line and tell you that my observations and knowledge of the Industry would indicate that all Retail milk Companies are passing through a most difficult time.

You no doubt appreciate the fact that during my association with you, the spreads were wider and general costs of operation lower.

I feel very strongly that if your association members will remember that tough days from time to time trouble us all, you are well on the way to helping a situation when it is recognized and has been tackled. You have in your favour both loyal and hard working associates, and with patience, in six months time I am sure you will find an improvement that will be most gratifying to your Board.

Sincerely yours,

Phil.

Philip Fleming.

PF/pl

Letter from Philip Fleming to W.W. Michell, Island Farms, 1947.

GEORGE AYLARD COLLECTION

Island FARM NEWS

Volume 1, No. 1   VICTORIA, BRITISH COLUMBIA   MARCH, 1947

Two New Directors Elected by Milk Co-Operative at Its Annual Meeting

THE ANNUAL GENERAL MEETING of the Island Farms Co-Operative Association, which was held in Victoria on the 28th of February, was well attended and was noteworthy for the determination expressed by the members to get behind the work of their large and growing Association. The meeting was presided over by Mr. W. W. Michell, president.

Two new directors were elected to the Board, Mr. Arthur W. Aylard replaces Mr. F. G. Waide, whose resignation had been accepted because he had sold his farm and was no longer an active shipper. The new director is the owner of Brackenhurst Jersey Farm, Sidney, V.I. He is a graduate in Agriculture of the University of British Columbia and has won national recognition for his fine, pure bred Jersey herd. He was recently awarded the title of "Constructive Breeder" by the Canadian Jersey Cattle Club. To win this certificate high standards must be attained; and for a herd to qualify, there must be an official production average of 400 pounds butterfat per cow in a herd of twenty-five cows or larger, or 425 pounds in a herd of ten to twenty cows, with classification average of 82.5 per cent of all cows in the herd. The average production of Mr. Aylard's herd was 477.53 of butterfat per cow under test, with a classified average of 84.22 per cent. Other requirements for the award state that the herd must be disease-free and every owner must have freed at least 65 per cent of all over that have freshened. All but one of Mr. Aylard's cows are of his own breeding. Mr. Aylard's 350-acre farm is the only dairy farm in Canada having raised and bred five cows that have produced more than 4,000 lbs. of butterfat.

Mr. John Martin, whose dairy farm is on the Sooke River at Milne's Landing, B.C., and who is Secretary of the Metchosin and District Farmers' Dairy Union, replaces Mr. Dan Chapman of Duncan. Mr. Chapman resigned from the Board due to pressure of his activities as Reeve of Cowichan. Members have received a printed copy of the annual report of the Directors and Auditor and the financial statement for the year ending December 31, 1946. This report showed the rapid growth of the Association and gave an optimistic picture of its future development. The assets of the Association have practically doubled during the past three years with substantial investments in new plants and equipment at Victoria, Duncan, Port Alberni, Qualicum and Parksville.

BUTTERFAT BONUS

It was pointed out in addition to the top market prices and other privileges, which members of the Association have enjoyed, that another bonus of 5c per lb. butterfat has been credited the members for the past year's operation. In addition it is indicated that a substantial cash payment will be made next year to members from the revolving loan fund.

LOYALTY URGED

Discussion at the meeting emphasized that the Association and other dairy organizations are selling more fluid milk and cream than ever before. For this reason it is easier for private dairy firms to play favoritism among shippers and to try to pull them away from their co-operatives. Members were plainspoken in urging loyalty to their co-operative marketing associations, which is their best protection in the face of the problems which will affect the dairy farmer in times ahead. The trend is more than ever towards co-operative marketing of primary products, and only by the continual strengthening of producer-owned organizations can farmers have the security and control necessary in their best interests.

B.C. FARM SURVEY

According to figures just released by the Dominion Bureau of Statistics, there are 26,372 farms in British Columbia.

The average number of livestock per farm is as follows: Horses 2.2, milch cows 4.6, other cattle 10.2, sheep 4.7, swine 2.6, poultry 172.8.

Higher Butter Prices

An increase in the price of butter effective May 1st is practically certain, according to reports from Ottawa.

The producers' subsidy of 10 cents per pound of butterfat used in the manufacture of creamery butter runs out April 30. Should the government decide to withdraw the subsidy on butterfat, a lifting of the ceiling would be necessary if the farmer is to receive the same amount for his butter that he now is. Actually, due to rising costs, dairy farmers are not receiving a fair price for their butter.

Recently, the Canadian Federation of Agriculture presented a request to the government for an increase in ceiling prices on dairy products to compensate producers of milk for increasing costs.

HAD TO IMPORT

They point out that ceiling prices are being maintained at such a low level that production is being discouraged, with the result that the government has had to import butter and has even considered the use of oleomargarine.

In respect to butter, the government can do one of three things:
1. Cut the subsidy, in line with government policy in respect to the eventual withdrawal of all subsidies, and raise the ceiling slightly.
2. Remove the subsidy entirely and raise the ceiling to compensate for the loss of the subsidy to the farmer and even provide some additional profit.
3. Remove the subsidy and the ceiling at the same time.

CO-OP BOOSTS CAPITAL

CRANBROOK—Capital increase of $10,000 has been approved by Cranbrook District Farmers' Co-Operative. It will be used to expand current operations.

One-fifth of the objective amount was pledged by members. H. C. King, manager, stated the excellent condition of the society's finances is due to the co-operation among the members.

ASSOCIATION DIRECTORS

The above photograph was taken following the Annual Meeting of the Island Farms Co-Operative Association. Seated from left to right are George W. Malcolm, Secretary and General Manager; W. W. Michell, Saanichton, President. Standing from left to right are John E. Martin, Sooke; A. W. Aylard, Sidney; Fred Wilson, Cedar; Robert Cheyne, Auditor; Harry Dawson, Nanoose; Capt. C. L. Anderson, Cobble Hill, Vice-President.

Island Farm News, March 1947.

GEORGE AYLARD COLLECTION

Events in Cedar

The Cedar Local of the
Island Farms Cooperative
Association meets the third
Tuesday in the month at 8:00
p.m. in the Community Hall.
A picnic is being planned by
the local for August 17th,
with Tom Michael to report
on arrangements at the next
meeting.

Island Farm News,
June 1947.

II — but I had met a Victoria girl and we had decided to get married and stay there. Although I was not strictly a founding employee of Island Farms, I joined as office manager early in 1945. Having recently been released from the Royal Canadian Navy after five years of service, I learned of the vacancy at Island Farms through a naval friend who had decided to make the navy his career and therefore he had turned down the job at Island Farms. At that time, the company was just getting started from a plant at 608 Broughton Street, Victoria. The building and adjacent lot are still there. The site was even then unsuitable, which led to many problems.

The general manager was Mr. Philip Fleming, an administrator and a social person; that is, he belonged to the Union Club, the Rotary Club, the Chamber of Commerce, etc. His assistant was George Malcolm. He was a hands-on dairyman who seemed to know every farmer on Vancouver Island who had at least two milk cows. The plant manager was Norman Ingledew, a University graduate in bacteriology. The sales manager was Thomas Harkness. George Malcolm was the organizer who drummed up enthusiasm among the farmers on the Island, and who persuaded them to become members of this new cooperative dairy firm. Our main opposition at that time was Northwestern Creamery, but Palm Dairy and Shepherd's Dairy were also active competitors. Most of our plant equipment in those early years was second-hand and had been purchased in various places on the mainland. As we grew from six or seven routes to eventually 20, we needed to purchase much of our milk as well.

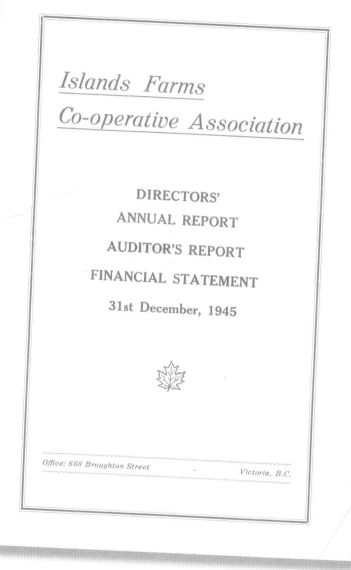

Island Farms Cooperative Association annual report, 1945.
GEORGE AYLARD COLLECTION

Christmas party, Island Farms, late 1940s.

Fluid Milk Rate Too Low

This is an application to the Milk Board by representatives of the following organizations:

Vancouver Island Dairymen's Association;

Island Farms Cooperative Association;

Sooke, Metchosin Farmers' Union

Mid-Island Dairymen's Association

and by representatives of individual primary producers not affiliated with any producer associations. The application is for an increase of 40 cents a hundred pounds in the rate paid for fluid milk.

The application to the Board for an increase in producer returns is based on the grounds that since 1942, there has been a steady increase in the cost of milk production without a corresponding increase in the price paid to producers.

The Board has, therefore, decided to grant the application of the producers and an order will issue establishing this price effective on and from April 16, 1947.

Excerpt from decision by Milk Board chairman, E.C. Carr dated April 16, 1947.

At that time the passenger boats were on regular runs to and from Vancouver. So we bought from 10 to 20 ten-gallon cans of milk every day from the Fraser Valley Milk Producers' Association. This milk arrived early every morning on the night boat and we picked it up in a beaten up old truck. All of our vehicles were old, and held together mainly by wire because no new ones had been built throughout World War II. So we began life with a collection of wrecks. At that time in 1945, we still had several routes being serviced by horse-drawn wagons and our barn was right in the centre of Victoria, approximately across the road from the present Hudson Bay parking lot, between Douglas and Quadra streets.

The entrance as I recall was on Johnson Street, not far off of Douglas. This property was rented, not owned, and I believe we got out of the horse business about 1947 as new trucks became available. We had a priority to purchase new trucks because of the nature of our business as the manufacture of new vehicles re-commenced after the war. Sales of new vehicles were very carefully monitored because the scarcity caused a flourishing secondary market on new vehicles, whether they were cars or trucks. Our general manager, Mr. Fleming, managed to get a new car on this priority basis, and it was quite a popular vehicle, believe me.

Returns from BC creameries, 1907.   BCARS GR 509

Name	Address	No. of patrons.	Lbs. of Butter	Gross Receipts	Sold per Lb.	Amount paid Patrons.
Chilliwack Cream. Co.	Chilliwack	150	256,000	$80,911.90	33½¢	$65,073.40
Edenbank "	Sardis	92	222,200	67,981.00	33	60,315.16
Cowichan "	Duncan	120	179,000	61,882.00	34.5	57,139.08
Victoria "	Victoria	113	180,930	64,128.00	35-4/10	55,649.86
Vancouver "	Vancouver	70	150,000	47,000.00	32	40,500.00
New Westminster "	New Westminster	75	133,285	44,669.00	33½	38,380.59
Delta "	Delta	40	97,500	38,000.00	34	30,000.00
Comox "	Courtney	31	78,782	27,290.00	34.64	23,634.60
Nanaimo "	Nanaimo	50	69,195	24,644.35	35½	21,502.45
Ford Saxon "	Vancouver	40	75,000	25,500.00	34	21,500.00
Sumas "	Upper Sumas	18	46,275	15,011.87	32½	11,509.11
Salt Spr. Island "	Ganges Harbour	32	41,488	14,328.66	34-5/10	11,492.20
Abbotsford "	Abbotsford	32	35,347	11,310.00	32	9,275.00
Glenbrook "	Mission Juct.	57	25,240	8,834.00	34.5	7,500.00
Alberni "	Alberni	22	18,600	5,901.00	31	5,035.00
White Valley "	Lumby	30	22,000	9,000.00	27	4,500.00
Okanagan "	Armstrong	17	17,422	4,979.31	28½	3,017.79
Peerless "	Abbotsford	15	3,000	1,050.35	35	800.00
Butter Total		994	1,651,304 lbs	$545,421.43	32-15/16¢	$466,824.23

			Lbs. of Cheese			
CHEESE FACTORIES						
British Columbia Cheese Co. Langley		20	95,000	12,800.00	13½	10,800.00
Companies handling milk only						
Richmond Dairy Co.	Vancouver	50	milk only			59,776.75
Almond Dairy Co.		40	" "			58,000.00

From 1945 to 1950 we expanded into Duncan, Qualicum, and Port Alberni. The first two were distributing points only, whereas Port Alberni was a full plant. In that area the only opposition initially was McKinnon Farms, brothers who sold bottled milk — no relation of mine. None of these remote stations were profitable or successful. So we closed Duncan eventually, sold Qualicum to H.R. MacMillan Farms, and disposed of Port Alberni as a plant. This was done after

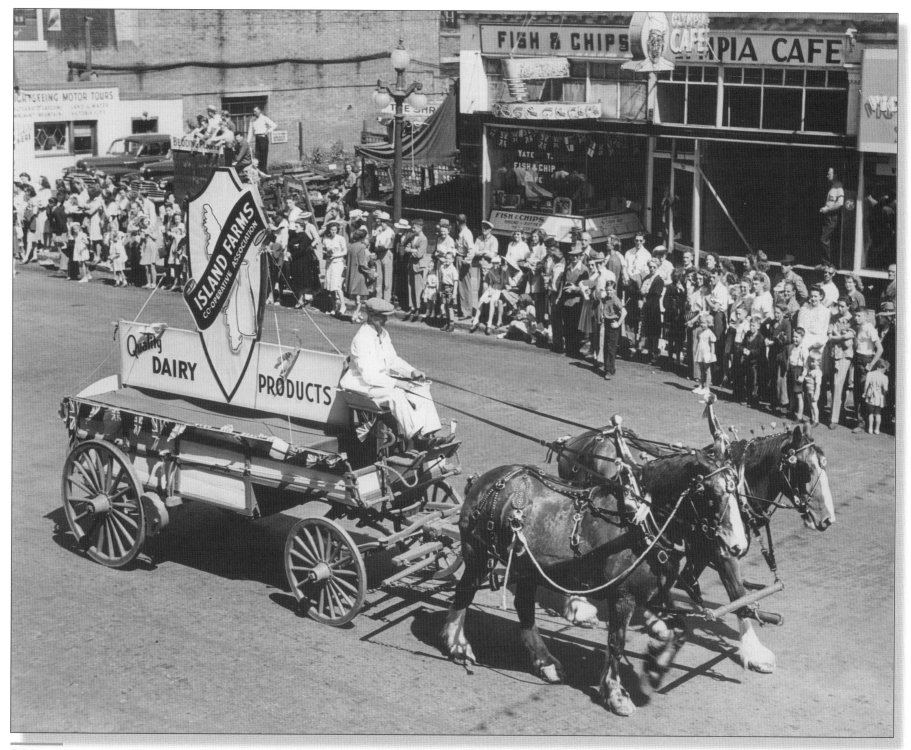

Ed Low driving Island Farms wagon in May 24th parade, Victoria, 1947.

100 Percent Turnout Is Essential

It is regrettable that the Board of Directors have found it necessary to make a deduction of 18 percent from your milk cheque in December.

Our present position is caused, to a large extent, by the fact that our members' production is at an all-time low. In December, for our Victoria plant, we purchased $14,107 worth of milk from members and $15,732 from outside sources. How long can we suffer from this unbalanced winter and spring production?

Our Board of Directors has decided to call our Annual General Meeting as early as possible in February to give the membership an early opportunity to discuss the affairs of the Association. A 100 percent turnout is essential at this meeting.

Letter to members from N.H. Ingledew, general manager of Island Farms, dated 19 January, 1949.

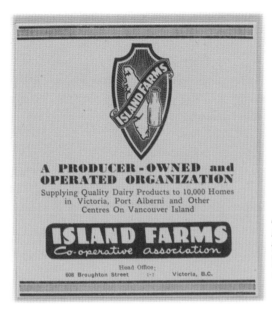

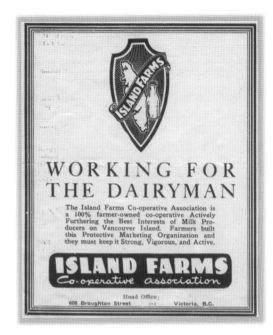

Advertisements for Island Farms in *The Farm News*, 15 July 1948.
GEORGE AYLARD COLLECTION

I left in 1951, so I'm not sure how the disposition of the Port Alberni plant occurred.

Cash was always a problem in the early days. Our broken-down plant equipment and trucks were costing a fortune to maintain and we didn't have funds for replacements even when new trucks were available for $800. That was the price of new trucks when they first came off the production lines in 1945. In addition, it was obvious that our downtown location would have to be abandoned soon. So about 1950, we started to look around for a better site.

A constant problem in the early days, especially when we didn't have enough cash from day to day, was our mounting accounts receivable. Our commercial customers — such as restaurants and corner stores — bought milk, cream, and ice cream, butter and eggs from us and would be billed at the end of each month. They would then wait another 30 days at least before considering paying us and suddenly they would owe 60 days and be into the 90 day period, and we simply could not afford this sort of a situation. Competition was fierce, so we really didn't want to lose them and yet

we were caught … Today they would call that being caught in a "Catch 22" situation. As a matter of fact our sales manager, Tommy Harkness, and myself — even when I was manager — we went out to collect money directly to stores and restaurants. I recall clearly for months on end he had three stores to go around to on Friday night or Saturday night, and I had two restaurants, and these restaurants were open until 1:00 in the morning. Now, in those days we didn't have any television, so you were hard pressed to stay up until 12:30 to get in your car and go down to restaurants and try to stand around, embarrassed of course, because you stood out, hovering around the cash register waiting for the manager or the person in charge to hand you a few dollars, five dollars, ten dollars, whatever they had left after the other creditors had been around too. I always had a receipt made out in advance, so that I could get out of there as quickly as possible. We kept this up for several months, and we had to give it up because it wasn't really worth the effort. But it was one of the interesting things that happened in the early days. ■

Island Farms Co-operative Association

Victoria, B. C.

Mr. E.A. Aitken,
Duncan, B.C.

Mr. W. Allen,
Coombs, B.C.

Mr. A.W. Aylard,
Sidney, B. C.

Mr. John W.C. Barclay,
Sidney, B. C.

Mr. Elton Bayne,
Alberni, B.C.

Mr. S.R.D. Bayne,
Alberni, B.C.

Mr. D.A. Beaton,
Qualicum Beach, V.I.

Mr. J. Bigmore,
Alberni, B.C.

Mr. A. H. Brittain,
Coombs, B.C.

Mr. C.P. Brittain,
Hilliers, B.C.

Mr. S. R. Brittain,
Hilliers, B.C.

Mr. T. C. Brock,
Brentwood Bay, B.C.

Mr. Richard Brown,
Cobble Hill, B.C.

Mr. Walter Burnside,
Duncan, B.C.

Mr. E. Campbell,
Royal Oak, B.C.

Mr. M. Casey,
Saanichton, B. C.

Chapman Bros.,
Duncan, B.C.

Mr. R.H. Chappell,
Sidney, B.C.

Mr. L. Comeau,
Victoria, B.C.

Mr. Robert H. Connell,
Victoria, B.C.

Mr. J.L. Cresswell,
Cowichan Station, B. C.

Mr. S.J. Darby,
Alberni, B.C.

Mr. H. Donald,
Sidney, B.C.

Mrs. A. Doran,
Victoria, B.C.

Mr. C.E. Doty,
Duncan, B.C.

Eaglecrest Estates Limited,
Qualicum Beach, V.I.,

Mr. William T. Easthom,
Qualicum Beach, V.I.,

Mr. A.W. Evans,
Duncan, B.C.

Mr. L. Watson Evans,
Duncan, B.C.

Flett Bros.,
Duncan, B.C.

Mr. J.H. Forge,
Sidney, B.C.

Mr. H. Greenard,
Alberni, B.C.

Mr. N. L. Grieve,
Royal Oak, V.I., B.C.

Mr. Eric Hamilton,
Duncan, B.C.

Mr. J. W. Hansen,
Errington, B.C.

Mr. M. Hansen,
Cowichan Station, B.C.

Mr. G.W. Herwaarden,
Alberni, B.C.

Mrs. A. Holman,
Westholme, V.I., B.C.

Mr. R.J. Horton,
Cobble Hill, B.C.

Mr. J. Houston,
Cliffside P.O., B.C.

Mr. S. Jackson,
Sidney, B.C.

Mr. Victor W. Jopp,
Victoria, B.C.

Mr. Jas. Kennedy,
Victoria, B. C.

Mr. Walter Kyle,
Alberni, B.C.

Mr. W. Laird,
Sidney, V.I., B.C.

Mr. C. K. Lang,
Royal Oak, V.I., B.C.

Mr. J. Looy,
Saanichton, V.I., B. C.

Mr. A. J. Lowery,
Royal Oak, V.I., B.C.

Mr. Chas. Lucas,
Cowichan Station, V.I., B.C.

Mr. Thomas H. Lunson,
Victoria, B.C.

Mr. Preston MacGowan,
Parksville, B.C.

Mr. Andrew R. McGregor,
Prospect Lake P.O., B.C.

Mr. Gilbert H. Mant,
Qualicum Beach, V.I., B.C.

Mr. J. E. Martin,
Milnes Landing, V.I., B.C.

Mr. J. Metcalfe,
Duncan, B. C.

Mr. W. W. Michell,
Saanichton, V.I., B.C.

Miss G.E. Moses,
Sidney, B. C.

Mr. M. Mosher,
Alberni, B. C.

Mr. John Musselwhite,
Parksville, B.C.

Mr. J. W. Mycock,
Errington, B. C.

Mr. N. H. Mycock,
Royal Oak, B.C.

Mr. M. C. Nissen,
Sooke P.O., B.C.

Mr. J. F. Nesbitt,
Cobble Hill, V.I., B.C.

Mr. H. Ottosen,
Cowichan Station, B. C.

Mrs. W.H. Ould,
Cobble Hill, B. C.

Mr. C.J. Owens,
Port Alberni, B. C.

Mr. S. Owens,
Cowichan Station, B.C.

Mr. Wilfred J. Palfrey,
Victoria, B.C.

Mr. F. R. Parr,
Cowichan Bay Road,
Cobble Hill, B. C.

Mr. L. D. Pickard,
Duncan, B. C.

Mr. C. Randall,
Errington, B.C.

Mr. T. Raper,
Errington, B. C.

Mr. C. Reimer,
Sidney, B. C.

Mr. E.W. Russell,
Qualicum Beach, B.C.

Mr. Dave Schochenmaier,
Sidney, B. C.

Mr. Alvyn Shannon,
Alberni, B. C.

Mr. E. W. Smith,
Coombs, B. C.

Mr. James B. Smith,
Coombs, B. C.

Mr. M. J. Smith,
Duncan, B. C.

Mr. J. Somers,
Alberni, B. C.

Mr. A. Spotts,
Royal Oak, B. C.

Mr. O. G. Strand,
Errington, B.C.

Mr. D. Thomas,
Saanichton, B. C.

Mr. W. A. Thompson,
Alberni, B. C.

Mrs. Esther Tryon,
Parksville, V.I., B.C.

Mrs. Nellie Volkman,
Alberni, B. C.

Mrs. D. Scott Wade,
Cobble Hill, B.C.

Mr. W. H. Wilkinson,
Cobble Hill, B.C.

Mr. F. Wilson,
Ladysmith, B. C.

Mr. Edgar J. Wood,
Parksville, B.C.

Island Farms Co-operative Association,
Port Alberni Branch.

Shippers list, Island Farms, early 1940s.   GEORGE AYLARD COLLECTION

CHAPTER IV

Serving Urban Markets and Distant Places

The *S.S. Rossland* at Arrow Park. Paddlewheelers like this one picked up milk cans and took them to creameries operating in the large centres of the Kootenays, 1906.
BCARS 02615

Delivery by sleigh, 1940.

Columbia Dairy, Trail

The Boumas have been running the Columbia Dairy in Trail for three generations. When Jan Bouma came from his native Holland in 1951, he worked for a few years in the Fraser Valley, then five years later he bought the Columbia Dairy from his uncle. Jan's uncle, August, came to the farm from Holland in the 1920s. At that time there were 4,000 people living in Trail and the town was growing because of the Cominco smelter. Since it was felt milk counteracted some of the harmful effects caused by overexposure to lead, for many years Cominco provided free milk to their employees, so the demand for milk was high. This is when August started up the dairy. It wasn't too long before he was joined by his brother, Jelle.

Each day one of the brothers took the milk into Trail — at first by horse and wagon and later by truck. Jan points out that in those days, there was no highway. The gravel road was rough, full of pot holes and "no wider than a kitchen table," he says. The day started early, particularly in summer when it was important for deliveries to be completed before the sun got too hot. A canvas soaked in cold water and draped over the milk in the wagon or truck helped to keep it cool. During bad winter storms in the days before snow plows, people would have to ride the roads all night so that the milk deliveries could get through in the morning.

From an article by Pam Humphreys, Butter-Fat, *October 1986.*

Moving milk in British Columbia has always been a difficult proposition because of the rugged terrain and long distances between towns. In the early years, the sticky gumbo of a muddy spring or the drifts of a big snowfall made road access to many places impossible for at least part of each year. Settlers first relied on waterways for movement and travelled up and down the coast, and on rivers and lakes by steamer and paddlewheeler. Later, the building of railways in the era of expansion between the 1880s and the mid-teens of the new century made possible a greater ease of movement for goods and for people. In this period, rail lines important to the development of the dairy industry were built, including the Canadian Pacific Railway (CPR), the Vancouver-Lulu Island, the Esquimalt and Nanaimo (E and N), the British Columbia Electric Railway (BCER), the Great Northern Railway (GNR), and the Canadian National Railway (CNR). Significantly, the building of railways to tie together different parts of the province not only aided in moving milk to market, but also helped to create markets by making the settlement of new areas of the province a possibility. Later still, the improvement and paving of roadways became a priority and goods and people gave up their reliance on the railways as they took to the highways. While new transportation corridors made possible an ease in the movement of people and goods, they also set the stage for new kinds of challenges as local balances between producers and distributors were upset.

The growing and changing infrastructure of road and railway in British Columbia was vital to the movement

Jim Casanave Sr. working for Prince Rupert Dairy, Prince Rupert, 1920.

Jim Casanave Sr., producer-vendor, delivering milk in Victoria, 1914-15.   JIM CASANAVE COLLECTION

of milk from farm to processing plant. But a network of people, horses, and eventually trucks took care of the movement of milk over the smaller, but no less vital, distance between processing plants and consumers at home. At one time, over 90 percent of fluid milk produced in the province was distributed through home delivery.

On the Milk Route
Frank Bradley, 2000.

> It is interesting in looking back to those horse and wagon days of the 1930s and 1940s to recall that in addition to the milkman there were many other people employed in home service with horses or trucks. The bread man for example would call about three times each week, the vegetable man probably twice, the laundryman once, the ice man maybe twice, coal came about once a week, the junk man would drive down your lane calling out for purchase of any junk you might have. He also had a horse and wagon. Department stores used teams of horses — a lighter type of horse — and the breweries also used teams of heavy horses, such as Clydesdales, to pull their heavy wagons. But most visible of all, even though most of the delivery work was done during the night, was the milkman, because the milk wagons or trucks were out on the road 365 days a year.
>
> If you were a producer-vendor or working for one, then you may have been involved in milking cows or washing bottles as well as delivering milk, and your

Mrs Kirby's dog cart, used for milk deliveries in Victoria in the 1930s.

Delivery in Victoria

In those days, there was no such thing as supermarkets. We had one wholesale route and I think the driver did his job in about four hours every day.

We had 20 routes. People put the milk bottles out on the front porch with a little note saying how much they wanted, unless they had a regular order. They were supposed to put their money in the bottle. Well, some of them did and some of them didn't. But if they didn't, they sometimes claimed that they had and it was the word of our driver against their word.

The horse would get to know the route just as well as the driver, and he would plod along and stop at a house he knew, and our drivers, who would have a couple of metal containers holding six or eight quarts of milk each in their hands, would be able to cut across from one house to the other without going back to the street. And the horse would plod along and stop at every one of the houses. It saved the driver going back and climbing in and out of a truck every few minutes. However, those days are pretty well gone. Today, with your huge supermarket activities, it takes large trucks to deliver these large quantities.

Donald McKinnon, 1993.

hours of work and your income would be dependent on the size and the profitability of the business. But if you were working for one of the established dairy processors, then your income as a milkman in those days would be about $65 per month, plus some commission on sales if your route was large enough to qualify. This in some cases might total as much as $100 per month.

You would work seven weeks straight and then get one week off. The work day would vary from 9 to as much as 12 hours depending on the number of customers on the route, the area covered, the weather and the number of collections or canvassing callbacks that you had to make, and, of course, dependent on the speed that you decided to travel. There was no payment for overtime, no statutory holidays, and if you were lucky you got one week annual vacation.

Today you might ask, "How did they find people who were willing to work those hours for those wages?" But you have to remember, those were the years of the Great Depression and there was no shortage of people looking for work. For example in 1936 a small three-line ad was placed in the help wanted section of The Vancouver Province *seeking applications for the position of a milk route driver and citing a box number at* The Province *for replies. I talked to the son of the owner of that small business 54 years later, and he told me that he always remembers going into the office of* The Province *to pick up the replies, and was amazed to find that there were over 200 of them.*

The thought of a regular paycheque was appealing to a lot of people in those years. In addition, even if you were working for one of the dairies, there was an aspect

Teamsters agreement with Fraser Valley Dairy, 1919.

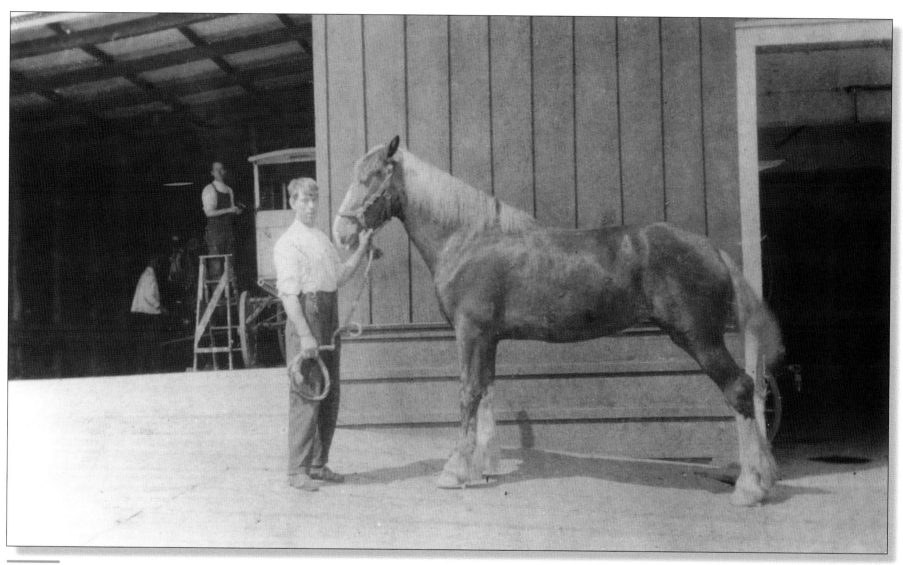

Horses at the FVMPA's 8th Avenue plant, 1920s.

Lake Windermere Creamery

Memo for the information of applicants for the post of manager:

The settlers in the Columbia Valley, in conjunction with a firm of bankers in London, England, who are financially interested in the Valley are about to erect a creamery etc. near Lake Windermere Station. The Committee of Settlers to whom preliminary arrangements have been entrusted are prepared to engage a manager provisionally at once.

The man appointed must be an experienced butter-maker, an organizer, a man of sound business methods and habits and of equally sound character, and possessed of tact, energy, and initiative. The salary offered is $150 per month until the company begins operations, and thereafter $175 per month.

Applications should be addressed to the Chairman of the Creamery Committee, Mr. James Sinclair, Invermere BC.

Notice dated June 25, 1922, from Dairy Branch papers housed at the British Columbia Archives.

of independence associated with being a milkman. In some respects it was almost like being in business for yourself. For example, on a horse and wagon route, you arrived at the barn about half an hour before your scheduled loading time, which might be two or three o'clock in the morning. The horse would be harnessed and ready for you, except for the bridle, and you hitched him up to the wagon, lit your kerosene lamp for the tail light, then off to the loading dock.

On leaving the dairy you were on your own. Whether you hustled around and finished the route in seven hours or took it easy and made it in nine hours, it was up to you. But you were expected to serve all the customers in a satisfactory manner so that there were no complaints. You had to keep in mind that some of your customers would expect you to be there before their breakfast time and all the customers liked to have a

Albert Ware, Jersey Farms blacksmith, 1940s. FRANK BRADLEY COLLECTION

delivery time that they could rely on, particularly during the warm weather. They did not want the milk left out on the doorstep too long before being placed in the ice box.

Each month you were required to write up a route book listing names, addresses, and methods of serving all your customers and this would be the book that you would mark the transactions for that customer, what products they bought, whether they paid or charged, and the number of bottles short or over — there was a five cent deposit charge for bottles. Your books had to balance when you turned the route over to a relief man on your days off, so any losses were your responsibility.

For customers who were charging their purchases, it was your responsibility to write up bills and make collections. This was a very important part of the job and often it was time-consuming too because you often had to call back on some customers who would not want to leave the money on the doorstep overnight.

Competition for new customers was keen, but once established with good service and good products, customers did not readily switch. A good milkman always had his eye out looking for new customers and the best opportunities were often when a family moved into a new home, so it was important that you keep a close eye on the empty houses on your route. Some companies had arrangements for advance information from moving companies or with real estate firms. Sometimes canvassers were employed to go door to door and the average pay used to be about one dollar for each new customer secured, but this was not always too successful as some canvassers would make

Morning milk train, Deroche, 1920s.

Jersey Farms wagon carrying Bud Morton and John Carlyle, undated. FRANK BRADLEY COLLECTION

some pretty wild statements in order to obtain customers. Sales contests and extra commission incentives were used to increase sales of products such as butter, eggs, and cottage cheese. Butter sales were much better than they are now because, of course, at that time there was no margarine on the market.

In some cases the resident caretakers in some of the apartment buildings were given a janitor's allowance, usually a pint of milk daily, in return for the exclusive right to serve all tenants. The average route would serve anywhere from 250 to 300 customers. On some routes, such as the West End where there were a lot of apartments, the number of customers would be much higher;

however, many of those customers were what is known as WBOs — you only served them when the bottle was out.

These apartments were built with a little cubbyhole at floor level with a door on either side opening into the hall and into the apartment, so you would sometimes travel down the hall, opening a lot of doors, but not finding too many bottles. The general procedure was that you would have two hand carriers with eight spaces in them for quarts, pints, half pints, etc. You loaded them up with what you expected you might sell in the apartment, and you proceeded to work from the top floor down. Invariably, someone would want something that you didn't have in the carrier and you would

Kitsilano depot of the FVMPA, late 1920s.

Morris Pihl

Homesteading one and one-half miles east of Aldergrove in 1907, Mr. Pihl cleared his farmland out of virgin timber and joined the FVMPA in 1917 when milk shipments began. He helped organize the Aldergrove Local and served as its secretary during the early 1930s. He was the first mail carrier in the Aldergrove area when rural delivery began in 1913.

The Pihl farm was logged with four-ox teams which hauled the timber over skid roads to a local sawmill. Mr. Pihl's son, Alfred, recalled using two cases of blasting powder to split one of the giant Douglas fir stumps that remained after the land was logged.

One tree was big enough to supply all the family's fuel needs for the entire year.

Butter-Fat, *January 1962.*

have to make a second trip. So I think that in the minds of most of the milkmen of the day, serving apartments was a bit of a pain in the neck, especially if they didn't happen to have an elevator.

As mentioned before, the milkman made deliveries every day — even Christmas day — and this created some problems. At Christmastime it was customary to reward the milkman for his service, and one of the most popular gifts in those days were tins of cigarettes (flat fifties). The milkman would come in at Christmas time with literally dozens of these flat tins of 50s along with a lot of other types of presents. However, the problem arose when some of the customers invited the milkman in to have a drink, and some were very insistent and it was difficult to refuse. Every time he did meant that he was getting later and later and that meant that the supervisors at the dairy, who couldn't go home until all drivers were in, would be lucky if they got home in time for Christmas dinner. The horse would get back to the dairy okay, even if the milkman was asleep in the wagon. I remember on one occasion the milkman came back riding the horse. Some Christmases were disrupted for the milkman and his family and for the supervisory staff who had to rescue him and finish the route. Eventually a joint public relations campaign by union and company in asking customers to refrain from pouring a drink was conducted to make the public aware of the problem and the problem disappeared.

The normal procedure with the larger dairies was to receive milk from the farm in the morning, pasteurize and bottle it that day, and ship it out the next morning. Any not shipped would go out the following day.

All unsold milk returned from the routes would be destined for the manufacture of ice cream or picked up by the pig farmers for pig feed.

Some of the dairies replaced their horse-drawn wagons with sleighs during the winter months if snow conditions prevailed. There was no snow clearing equipment at that time and only intersections on main streets were cleared by crews shovelling the snow into trucks for removal. Those kind of conditions caused customers to really appreciate the service they received when the milkman struggled through. During the night and on weekends, of course, he was usually the first one on the road and had to break trail.

I talked earlier about customer loyalty and I was thinking my own mother was a good example of this sort of thing. We started with Associated Dairies at our home in South Vancouver. A year or so later we moved to the Fairview area and the dairy followed. Then after three or four years there, we moved to the east end of Vancouver, 2nd and Nanaimo, and Associated Dairies followed us to that location. My mother and dad then moved to a few miles away to College Street and still used Associated Dairies, and then on to Oakridge; same dairy. Finally, when my mother went into a nursing home she had been a customer of Associated Dairies for I would guess about 60 years, and for 35 or 40 of those years I was working for another dairy, but she would never change and who was I to argue with that kind of loyalty. Ironically, when she went into the nursing home, guess who had the contract to serve it? Associated Dairies, now called Dairyland, of course.

Morris Pihl, dairy farmer, supplements his income by unloading shingle bolts beside the BC Electric Railway where it crossed Jackman Road near Aldergrove, 1913.

When Milk was Hauled by Mules

The last of January's snow was melting by the roadside. Frank Merson looked out the window of his Rosedale home and laughed.

"Everyone's going around with long faces complaining about this cold weather and they don't know how well off they really are. I can remember winters — lots of them — so cold the Fraser froze solid and you could cross with a team to Agassiz. Many's the time I hauled milk when the drifts were up to the roof of my garage."

On matters such as these, Frank Merson speaks from experience; he was a contract milk hauler in the Rosedale area for 49 years. He sold last summer [in 1961].

"I started hauling with horses in 1912, the year after I came from England ... I picked up cans along Yale Road and took them over to Edenbank Creamery. When it closed I hauled to Chilliwack Creamery and Borden's until the Sardis Plant opened in 1927 I think it was. I've been contracting for the FVMPA ever since.

Frank Merson pointed to a photo of a team of mules, one gray and one black, on his kitchen wall and explained how he came by them:

"The Army came to me in 1915 wanting to buy my team of horses. They needed horses for the war. I told them they could have the horses and I started looking around for something to replace them. There were a few mules in the area then and I had a couple of unbroken ones offered me. I bought them and set about breaking them. First time I had the gray one in harness she took fright and galloped all around Rosedale. I just let her go because I knew she'd be broke when she stopped. She was, too. Those mules were worth the work of breaking them in. They stood the road better — a horse needs a day's rest for every two on the road, but a mule can go every day — and they were smarter."

Frank Merson chuckled as he recounted how his mules went on to the plant without him one morning when he stopped for coffee. "By the time I caught up with the team at the plant, the boys had the wagon unloaded and I didn't have to roll a single can. I guess I should have made a habit of it."

In the early 1920s Frank Merson yielded to the influence of modernization and Henry Ford and put a truck on the road. The mules hung on for a few more years, then trucks took over altogether. "Mules were slow," he said comparing mules and horsepower, "but they could go in deep snow when trucks were useless. With the team I picked up the first cans at 7:00 a.m. and was home by 5:00 p.m. if the weather was good, or close to midnight if it wasn't. Of course, we couldn't keep the milk cool or get it to the plant as fast as today. Still, it got there in pretty good shape — except for one incident I remember. A farmer just down the way here forgot to put one of his can lids on tight when he put the milk on the stand at night. The plant sent the milk back. Like any farmer, he didn't like having his milk returned, but he could hardly say it smelled all right. A skunk had fallen into the can and drowned."

An interview with Frank Merson, printed in Butter-Fat, *February 1962.*

He Can't Win

In this passage from a 1950 Dairyland publication, Grace Luckhart reports on the challenges of the milkman's day.

"Oh, that's just the milkman," you say to yourself when you've identified sounds at the back door. Well, he may be "just the milkman" to you, but believe me, you're a lot more than that to him. You may not know it, but you're down in his book in black and white. A regular case history. Not for anybody to read — just for the Dairyland milkmen who serve the route on which you live. There is your own milkman who works eight days at a stretch, his relief man who takes over for two days, and the inspector who may appear at any time.

If you happen to live in the West End where the routes are heaviest, you will likely have a set of initials after your name. "Mrs. Smith, WBO." Don't be upset, or unduly elated about this. Because you and the 400-odd customers on your route nearly all have the same decoration. If you aren't a WBO, which in the milkman's language means deliver "When Bottle Out," you may be an EOD. That means deliver "Every Other Day."

In addition to the above initials which help him to interpret your wants, he also has to know where you leave your bottle. So he adds the initials "B," "F," or "S." That means deliver at the Back, Front, or Side door. In his book, he also jots down what you usually order — which and how many of the 12 items with which the Dairyland milkman stocks his wagon. Naturally he has learned to be a good guesser and more or less of a mind reader, but he isn't that good, even if eventually he gets to know you like a book. But the

Island Dairy, owned and operated by M. McNair, circa 1914.
ARCHIE MCNAIR COLLECTION

M. McNAIR, Prop.   Phone: Eburne 216-L-1

PURITY AND CLEANLINESS GUARANTEED

ISLAND DAIRY

Milk Fresh from Local Ranches on Lulu Island

SOUTH VANCOUVER P.O.

Business card, M. McNair, circa 1914.
ARCHIE MCNAIR COLLECTION

Hauling milk cans in Chilliwack, circa 1920.
ALLAN TOOP COLLECTION

Serving Routes during the Second World War

The following is a conversation, dated 1993, between Frank Bradley, Bill Osborne, and Jeff Harbottle. Harbottle's Dairy was a jobber of Jersey Farms operating in North Vancouver. It was eventually sold to Jersey Farms, where both Frank Bradley and Bill Osborne worked.

Jeff Harbottle and Frank Bradley beside Harbottle's restored dairy truck, a 1938 Diamond "T" one ton pickup, 2000. Bradley was its first driver on a home delivery route with average loads of 25 cases of milk. The truck is now owned by the North Vancouver Museum. NEIL GRAY COLLECTION

JEFF HARBOTTLE: If a truck broke down we would take the back seat out of the Chrysler and deliver with it, especially up North Lonsdale if there was heavy snow. That car with chains could go through snow right up to its bumper. My dad had a 1929 Desoto, with six wire wheels. He bought that car without saying anything to my mother. After he sold it we had a variety of trade-ins. One was a big Nash touring car with an overhead valve engine and we delivered in that. We only had two cases to start off with.

BILL OSBORNE: We had a couple of old REO speed wagons. The engine had a three-inch piston and a tremendous long stroke, opposite to what they have today.

FRANK BRADLEY: Do you remember when gasoline was rationed? That was about 1939 or 1940. You impressed on me about the gasoline and your theory was that we shut off the key and let the truck coast into a stop. I got so that I was pretty good at shutting the engine off and coasting. And it is amazing the amount of gasoline you saved. I remember one time about 2 a.m. when I coasted to a stop behind a car parked in front of my customer's house. I got out, got my two quarts of milk, went through the gate, left my milk on the porch, and was returning to the truck when the parked car was started up and drove away. A young lady opened the gate and ran up the path toward my flashlight, calling out "I'm coming, Daddy." I'm not sure whether she was more relieved or embarrassed when I told her I was the milkman, not her father. The incident brightened up a dull morning for me.

BILL OSBORNE: We carried on with that practice at Jersey Farms. We used to tell the drivers it was illegal to keep the trucks running and we had charts showing the fuel savings. The wear and tear on the starter wasn't too great. It was the ring gear. But we got to the stage that in the Divco we had a switch that was pulled out half way for the ignition and the rest of the way for the switch. We mounted those by the steering wheel, right near a person's hand. We did this with all the home delivery trucks so it was convenient for the driver to step in, pull the switch and be going. Now of course all the trucks at Dairyland are propane or diesel.

relief milkman hasn't the same opportunity to know you and your wants, much less the inspector. Hence the written record.

Actually, a good milkman could qualify as a weight-lifter, a bookkeeper, a credit man, a collector, a mechanic, and a mind reader. To say nothing of being a diplomat.

If you don't believe the latter, think of what he has to contend with. In the first place, he's working on a fairly tight schedule. In order to start delivery at 6:00 a.m., he has to be up at 4:30 so he'll have time to get breakfast, get to the plant, check over his load, and get started on time. His average route for a day has 250 calls. Think that over! In the West End where there are so many apartment dwellers, his calls average 400 a day. Of course, that's where the "EOD" people live. But in order to be sure he misses nobody, he has to open all the apartment lockers to see if the customer wants anything. Sometimes he opens 10 lockers before he strikes pay dirt.

With a schedule such as he has laid out, he can't stand around just admiring the flowers or chatting with a customer. Neither can he abruptly leave her. So what does he do? By the time he's out at his wagon, he's out of earshot, and the status quo has been maintained. But don't let anybody tell you that doesn't take some diplomacy.

Then there's the matter of the bottles, or the tickets. Some customers go along on the assumption that the dairy washes the bottles in any case. So they do. (They not only do that but they sterilize them like nobody's business.) Sometimes customers even leave milk in them

Bill Fowler, Jersey Farms delivery man, Vancouver, 1940.

Losing Good Men

It's a cinch now — compared to the old days. Used to lose some good men over nothing more than poor milk. They'd get so fed up that they'd quit and go for a job where they were closer to fresh supplies. Used to feel like quitting myself sometimes, trying to make desserts without fresh milk, but now ... we get it from Vancouver twice a week by boat. Our logging train brings it up from the beach and we keep it at proper temperature in our own refrigeration plant.

Unidentified logging camp cook, quoted by Mamie Moloney in "Beyond the City Limits," 1950.

and just dump the money in the milk! (*I often wonder what initials the milkmen put after that kind of woman?*) *It isn't the easiest thing in the world to dry out the bottle after you have washed it, but it would make the milkman's life a lot happier if you would just turn the bottle upside down for a few minutes so that the tickets don't drown when you drop them in. Of course, he'd rather you put the tickets beneath the bottle, or in the neck. It isn't much fun fishing for money or tickets in a wet bottle.*

Some of the notes left by customers for the milkmen are really collector's items. Here is one, which bears repeating:

Dear Milkman: When you leave my milk, knock on my bedroom window and wake me. I want you to give me a hand to turn my mattress. P.S. Hope you don't mind.

Another customer made this simple request:
My back door is open. Please put the milk in refrigerator, get money out of cup in drawer and leave change on the kitchen table in pennies because we want to play bingo tonite.

A dog owner wrote:
I'm sorry about the dog-bite, but he'll get to know you after a few months like he got to know the poor gas man.

The customer isn't all the milkman has to worry about. There are all sorts of hurdles to overcome. There is the weather, for instance. Just take the winter when our usual British Columbia climate disappeared completely, and we were treated to a good old-fashioned Eastern Ontario winter. Snowdrifts, icy slippery streets, tickets and money frozen to the bottom of the bottles. But you got your milk just the same.

Get past the weather and you find here and there a vicious dog that never gets used to the milkman ... The milkman may want to kick the dog right around the block, but can he do it? He cannot. He has his customer to think about. So he talks nicely to it and tries in every way to sell himself to the dog, hoping against hope that it won't tear his pants right off.

In the milk business, the customer is always right. If she runs out of milk some morning, and is waiting for it, the milkman is late. He may actually be right on the dot, but to her, he's late. If, on the other hand, she forgets to put out her bottles and the milkman assumes she doesn't want any milk today, she blames him for this too. He came too early.

He can't win.

Hauling Milk

In the following section, Lloyd Duggan, owner of Okanagan Dairy Transport and Kootenay Dairy Transport, describes his experiences moving milk to dairy plants.

My dad hauled milk cans starting in about 1924. I got my licence to drive when I was about 15 years old. I was driving for about four years before that delivering empty cans. We were located in Winfield and most of the milk went into Kelowna Creamery, or Tutt's Dairy

A 1932 GMC, the first truck used for milk pick up by Willoughby Rooke, Langley, 1935.

Bob Simpson on one of the first milk hauling trucks, a gas powered flat-deck Day Elder with chain drive and solid rubber tires on the rear axle, Chilliwack, 1920s. Later, the Simpson trucks were International Harvesters. His sons, Bill, Josh, and Wes and his daughter, Roberta, all drove the trucks.

Nelson Griffiths, first truck driver for the Okanagan Valley Creamery Association, circa 1927.

Advertisement for Hayes Manufacturing Company, 1930.

Walter Adams

During the mid-twenties, he was hauling milk out of Matsqui Prairie with a one-ton Chevrolet. A day's haul was 100 cans and it took three trips to cover the route. It was not an easy life. For five years Walter worked seven days a week — never taking a day off. The year 1926 saw him form a partnership with Tom Rottluff, member of a well-known Matsqui family. His brother Gordon bought out Rottluff in 1946 and the business became the Adams Brothers Truck Line.

Farmers served by the line will recall the yeoman service done by the Adams Brothers trucks during the 1948 Fraser Valley flood. The normal 34 tons of milk hauled daily dropped to 20 tons, but their trucks were out night and day doing the best possible job. One truck was used exclusively for hauling cattle off flooded farms.

Nearly thirty years of trucking has not slowed down this man. He is on the job every day. As Matsqui shippers can vouch, he has served them well.

Butter-Fat, *May 1954.*

Hauling milk to the FVMPA's Sardis plant, 1925.　　　DAIRYWORLD COLLECTION

as we called it in those days. We would pick up the cans from the farmers in the morning, go in and unload them, wash them, and deliver the cans back to the farmers in the afternoon. As I grew older, I was able to haul and lift the 10-gallon cans and load them. At that time, in the early 1950s, there were a lot of Dutch people coming in from Holland and they were very good dairy farmers. The population was really starting to grow in the Okanagan. There was a terrific influx of people. So you know, the whole thing really expanded very quickly.

In 1960 Shannon Dairy on Annacis Island was supplying product to their outlet in Kelowna, the Kelowna Creamery, so I used to work with them on week-ends and hauling milk and cottage cheese up to the Okanagan and ice cream back. At that time, Neil Gray was the Manager of Shannon (he eventually became CEO of Dairyworld) and Roy Dinsmore was the Plant Foreman. Roy's brother Bill also worked in the plant at that time. The Dinsmores — Roy, Bill, and Stan — were prominent names in the industry. In 1964, Shannon Dairy was merged with Dairyland. NOCA took over the farmers so then all the milk had to go to Vernon.

It was kind of ironic because when the milk was going into Kelowna, I'd get to the first farmer at 7:00 a.m.

and I'd have to wait for him to finish milking every morning. Then when the Kelowna Creamery closed, he was the last one on the line because we started at the other end, south part of the valley, and I'd get there about 11:00 and guess what? I had to wait for him to finish milking. It became very evident that the milk truck really dictated the lives of a lot of farmers and people. I mean everything revolved around what time the milk truck came. You waited 15 minutes at 7:00 and you waited 15 minutes at 11:00. Everything was run according to when the milk truck came.

There were a lot of interesting stories. One of the farmers one morning came out to talk to me. This fellow had a daughter and she wasn't too glamorous looking, especially with rubber boots on and just coming out of the cow barn, and that's when I used to see her. One morning this father came to me and said, "You know my daughter don't you?" I said, "Yeah." He said, "How do you like my farm?" I said, "Oh, it's a good farm." He said, "Well look, if you marry my daughter, I'll give you the farm…" However, I looked at the daughter and I looked at the farm, and I didn't think there was any even saw-off there, so I retreated quite quickly! These are just some of the funny things that happened along the way in the years gone by.

Standard Milk delivery truck, Vancouver 1919.

Cream Shippers

The top ten cream shippers to NOCA Dairy during August were as follows:

1. F. Peacock, Grindrod.
2. B. and U. Schilling, Darfield.
3. H. Naylor, Deep Creek.
4. N.A. Gillis, Falkland.
5. F.A. Dangel, Grindrod.
6. Mrs. V. Mazur, Enderby.
7. K. Rainer, Darfield.
8. J.W. Emeny, Enderby.
9. J. Seelback, Darfield.
10. W. Koersen, Enderby.

These top ten shippers produced a total of 4,490 pounds of butterfat during the month of August, averaging 449 pounds per farm.

The Cream Collector, *September 1966.*

You saw people in their good moods and when they were cranky and you just kind of felt the vibes in the air when you drove in the yard. I don't think there's any other occupation where you get to know families like you do hauling milk. They get to know the milk hauler well, too, and he's almost like part of the family.

As a milk hauler, you got so involved with every family because you watched the kids grow up. The kids would come out to the milk house, so I'd say, "Oh, Jimmy, how are you doing?" He'd be maybe four or five. I'd say, "Stand up against the milk house wall here, I'm going to measure you." I would put a little mark of pencil on top of his head and put the date down. I used to do that with 20 to 30 farmers' kids and it was kind of funny because the years went by, and I kept measuring little Jimmy or little Jane as they grew up a little bit more and a little bit more. Of course, the Dairy Inspector would come along and say they had to repaint their milk house. But inevitably they would paint around those letters, and some of them had been there for 20 years and some of the kids went to university and came back and actually really got a kick out of looking at these little lines. I've done it with my own grand kids as well.

After the Kelowna Creamery closed, NOCA took over and I took over all the hauling in the South Okanagan, and it was all in cans. Then in 1963, in March, we bought our first tank, a Delaval tank on an International truck, and it held about 3,500 US gallons which would be about 13,000 litres. In those days, it wasn't metric, but was in pounds and gallons. That truck and tank brand new cost $23,000. The hauling

First receipt issued by Lloyd Duggan for bulk tank pick up, Vernon, 1963.

Milk hauling truck owned by Archie Edmonson, circa 1910.

Every Day

Milk trucking is a 365-day a year proposition.
Austin Loney, milk hauler, 1950.

Wesley Simpson leaning against one of his family's International Harvester trucks used for milk pick-up. This truck had a pony axle behind the drive axle and hauled double-decked ten-gallon cans to the FVMPA's Sardis utility plant or to the Borden plant in South Sumas. Undated.

ALLAN TOOP COLLECTION

rate at that time was basically the same as what it is today, but of course now we go and haul 40,000 litres instead of 13,000. The stops are a lot bigger, which has sort of made everything more economical to operate. Then we started in 1963 with the first tanker, then we bought another one in 1964 and started hauling milk up to Prince George to Northern Dairies.

The milk was from Abbotsford and the Okanagan both because now we're under Milk Board jurisdiction. If you could move milk categorized as surplus milk to some place where they got Class 1, the farmer got a better return. Let's say the farmer got $10 more for Class 1 than surplus and it cost you $2 to haul it, then the farmers got an $8 benefit. So, that was the basis of what the Milk Board was supposed to do. Of course, at that time there were different areas. There was Vancouver Island, Lower Mainland, Thompson-Okanagan. These areas eventually amalgamated. Today they've even seen amalgamation between provinces.

In Kamloops there was a family by the name of Schrauwen and they had had a dairy there for a long

time. And then in the 1960s, the Armstrong cheese plant went broke, they took it over and they operated under Armstrong Cheese or Dutch Dairies. Then there was another farmer that moved to Sicamous, DeWitt. So some of those people who started their own dairy farms or dairy outlets are still operating, like the Blackwell Dairy in Kamloops.

In that time frame, say from 1960 to now, we've gone from one milk hauling unit to about 50 trailers hauling milk. We used to think 30 miles was a long way to go and now we pick up milk in Smithers some days and haul it all the way to Abbotsford and that's 850 miles. That probably is the longest distance in North America that anybody picks up milk. McBride milk all comes to Abbotsford. Creston milk all comes to Abbotsford. It was unheard of in those days; it was almost insurmountable to even think about it. It's such a change that young people can hardly conceive of what has happened in a scant 30 years. More happens now in a month than what used to happen in a year.

In terms of hecto-litres per man-hour in picking up

Arctic Ice Cream truck, circa 1920.

Skating

In the days before bridges, Barnston Island Farmers used a ferry to bring their milk across the Fraser. Pick-up was made at Port Kells. And they had real winters then. When the river froze over, the milk was skated across the ice on sleighs.

Butter-Fat, *1950s.*

Unloading cans, Chilliwack Cartage, Chilliwack, undated. Chilliwack cartage operated International Harvester single and tandem-drive trucks. All were covered vans, affording the milk some protection from the weather.

ALLAN TOOP COLLECTION

milk — a large farmer one time would be 500 litres. Now an average farmer is 5,000 litres. There were over 100 cream shippers in the Kelowna area at one time and 56 dairy farmers. That would be in the early 1960s, or the late 1950s. The whole milk industry in the Okanagan sort of shifted north because of the Kelowna area. How could you ever make enough money on a piece of pasture that you could sell for $10,000 an acre. I mean, economically, it just didn't work. Before the land commission came in, farmers sub-divided. Where the shopping centres are in Kelowna now, that was all dairy farms. I used to go down the main road and drop half-cream cans on the seat beside me and I got real slick at that. I'd pull over on the wrong side of the road, open the door, and throw the can, spinning it so it wouldn't upset, then I'd drop down a gear and away I'd go across the road — that's all the traffic there was. Now, when you look at it, it's the main highway.

It isn't that long ago. I guess the scary part about it is when you see what has happened in the last 30 years, you sort of program that into what may happen in the next 30. Your wildest imagination could become a reality. Who knows what's next. Dairying will always be important, but I think it's become very competitive because of the superstores and so on. We sort of drifted into a time where there were a lot of regulations. We had inspectors everywhere. A lot of it was good; we definitely improved the product. Then in later years we got into situations where the rules were still there, but they started doing away with the inspectors as they were cut back. Some of it I guess was good, but I think it deteriorated some of the qualities of the dairy industry that we created. The number of dairies has diminished. Like I said, there were 22 dairy outlets in 1956, I believe. Now we're down to two or three in the Okanagan. It's put more pressure on the farmers because they don't have as many places to ship. They had

First Electric Train,
Westminster to Langley,
July 1st, 1910

The BC Electric Railway Company's first electric train, 1910.

Last Can Hauler

Jim McDonald Jr. was the last can hauler for the FVMPA. His father, Jim McDonald Sr., started hauling milk in cans from the Mount Lehman area in 1929. Jim Jr. started working for his dad in 1940 and bought the route in 1953. He picked up milk from the Bradner-Mount Lehman area until 1966 when his territory expanded west to the Port Mann Bridge area and took in Barnston Island.

By the late 1960s, most of the larger producers in the Valley had changed to tanks, leaving a few of the smaller producers still shipping in cans. From March 1969 to February 28, 1970, Jim was the only can hauler remaining. In the mid-1950s, there were approximately 34 separate can hauling contracts to pick up milk for the FVMPA in the Fraser Valley.

Gerry Adams, 2000.

Milk Delivery, Island Farms, Victoria, 1949. Photo by Duncan MacPhail.

BCARS I-02199

to join together with the co-ops for protection and stability. Of course, you get into times where now we're getting the competition from foreign companies that sell truck loads of milk. They can make it pretty tough for co-ops. I believe in the future that there will be more farmers that will specialize. I can see that happening now, or at least starting to happen. There have been a couple that started and failed because they didn't realize all the marketing problems that they might run into. But these people are very intelligent and if they make mistakes the first time around, eventually they don't make any more. I'd be interested to see how things unfold in the next five years, but I'm sure it's like history. The egg rolls over and you start all over again. Everything we do is cyclical.

There's all kinds of stories about NOCA too. They were sort of a thorn in Dairyland's side for so many years. They would take a little milk down to the coast and sell it and then as the roads got better, Dairyland sent milk up to the Okanagan to their superstores. When NOCA was first conceived, it was North Okanagan Creamery Association, and then it was SODICA, Shuswap Okanagan Dairy Industries Cooperative Association. Originally the dairy was owned by Burns Brothers and they hired this fellow named Everard Clark, a young fellow, a real tyrant, to run it. He lived in an era of time where his management style was successful where others failed. Had he lived 30 years earlier or 30 years later, he may have failed miserably. But in the early 30s, things were tough. Co-ops

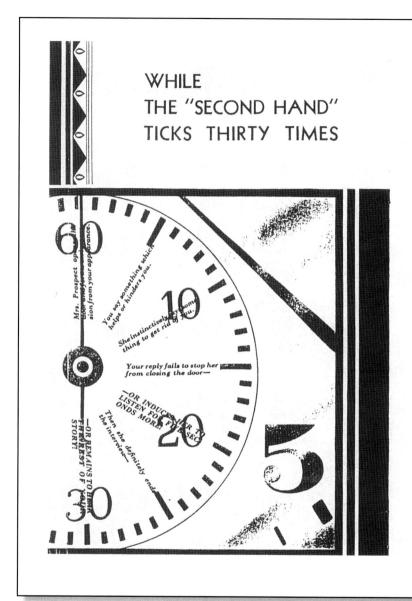

WHILE THE "SECOND HAND" TICKS THIRTY TIMES

Mrs. Prospect opens the door. She judges you and forms an impression from your appearance.

You say something which helps or hinders you.

She instinctively wants something to get rid of you.

Your reply fails to stop her from closing the door—

—OR INDUCES HER TO LISTEN FOR FEW SECONDS MORE.

Then, she definitely ends the interview—

—OR REMAINS TO HEAR THE REST OF YOUR STORY!

First Impressions Last!

LIKE that obstinate juror, Mrs. Prospect can't see things any way except her own! A salesman may be the world and all to his mother; his company may be the finest in its field; his products or service A-1 plus; but when Mrs. Prospect opens the door, all she knows about him is what she sees.

She won't ignore a slovenly appearance and look for the *"gold beneath the surface."* Nor will she think: "He doesn't look so '*hot*' but maybe he represents a fine company."

Instead, her impression of his products or service, will be only as good as her impression of the man she sees.

How important it is then, for the Route Salesman who meets her eyes as she opens her door—to represent himself, his company and what he is selling, to the greatest advantage. A clean appearance, well-cared for clothing, and a pleasing expression . . . details, yes . . . but vital matters in that first half minute.

Next in importance is what he says and how he says it. That will confirm the prospect's first impression or change it for *better* or for *worse*.

Her attitude will range anywhere from antagonism to eagerness to listen further.

Page from a sales book called *Illustrated Route Selling,* **Associated Dairies, 1930.**

Anyox, BC

Away up on the northern coast of British Columbia, some 650 miles from Vancouver city, lies the mining and smelting centre of Anyox. It is the most distant point that the product of our association in fluid form is shipped and from the time our milk leaves the farm in the Fraser Valley until it is delivered to the consumer in Anyox some three days must necessarily elapse. This calls for milk of a high-grade quality and as it must keep sweet for another three days, or until the arrival of another shipment by boat.

Now the people of Anyox, of whom there are over 2000, do not need to be told the value of milk as a health food, for dairy products are more of a necessity there than elsewhere, in order to counteract the fumes of sulphur and gas from the smelter, which kills all vegetation and growth for miles around. The product of the cow is used there extensively in the form of fresh milk, evaporated milk and reconstituted milk (made from milk powder), therefore "Fraser Valley" products, including butter, are very much in evidence in this northern outpost of our province.

Butter-Fat, *October 1930.*

were failing and because of his demands on people, he made the co-op survive because he made everybody working for him responsible for what they spent. He only paid them so much money per month, they would work on a commission or whatever it was, and he used to do all kinds of funny things. I remember when he had somebody out there delivering milk, the guy would run into the store or into the house to deliver, Clarke would sneak in the back and take out three pounds of butter or something and then hide it and see how the driver was covering up when he came in. He did all these little kinds of things in a way, I guess he did very well the first three or four years when Burns owned it, and they thought he was making too much money cause he had signed some kind of contract and they got rid of him. Well anyway, things started to fail again, so then the co-op had no alternative but to hire him back, he wrote his own ticket as to what he was gonna make. It included a percentage of all ice cream and the milk.

One time when I first started hauling bulk milk there were about nine other milk haulers in the Okanagan, all hauling cans. Nobody else wanted to go on the hook and buy a tanker, that was a lot of money in those days. I remember when I started I had to spend $22,000 to buy two trucks to haul canned milk, and I had $4,000 saved up to put down, and the whole first year's revenue was only $22,000. When you use those figures today, they sure don't fit.

There were times when we used to haul milk over to Golden to Dominion Dairies, through the Milk Board, because that was a deficient dairy and the Okanagan had surplus and the Kootenays had a deficiency. Of

SODICA's refrigerated delivery van, undated.
LLOYD DUGGAN COLLECTION

course they're dependent on milk, and when the roads are all shut down you can't go. We actually loaded the tanker on a flat car on a train and hauled it over. Then we used to haul milk to Creston over the Salmo-Creston highway to Nelson and when the roads were shut down, we'd have to go way down almost to Spokane and then come back up. We could hardly make that trip in time to get back to pick up milk the next day again. The farmers in the Kootenay area soon became very dependent on the milk haulers. We were almost like one big family, and it was sort of like we always stayed together. There were times when I'd say to them, "Look, the price isn't going to go up, but you have to have a tank that's large enough to hold six milkings because of the terrain. We need the extra time to get there sometimes." I never recall one farmer who argued about it. I mean there were times when even at six milkings, because of road conditions, we would get there and they would have the tank right full and the cows all in the barn ready to go. We have hauled milk in the Kootenays for 30 years now and there are fewer farmers, but there is three times as much milk.

When we first started hauling out of the Creston

Dairyland tank truck, early 1950s.

Helping Out

Back in 1935, one truck went off the road on Barnston Island. During the 1948 flood, trucks were going night and day helping the farmers. But it took 31 years of hauling milk before a load was lost. Local shippers quickly rallied around with horses, tractors and manpower to get the trucks rolling. We know all the farmers pretty intimately — there is always good fellowship — we are never in difficulty for long when in trouble.

J.A. Loney, Milk Hauler, Butter-Fat, 1950s.

area, slides would come down, and that created a lot of havoc for us. I don't know if you can realize the power of some of those slides. I've seen them just ahead of me. First of all, you see just kind of like a white cloud, it looks like a white cloud running down the hill, until all of a sudden you see big tree stumps coming up out of the fog, you know, like they're rolling. And then the whole thing comes down and hits the road like the old metal guard rails they used to have. They would just wind those up like a corkscrew and every once in a while you'd see one pop out of the slide. They come right down over you, right past you. It sure gets your attention. Then one time I got hit at Three Valley Gap. The snow moved the truck over about 75 feet, and it lifted the truck with a load on five feet off the blacktop because of the packed snow going under. It covered up the passenger windows and the side windows. The only way I could get out was to wind the windows down and climb out the window and it just so happened where the slide came down was where there was a pull-out next to the lake. A hundred feet either way I would have ended up in the lake.

Another time I was coming down the Salmo-Creston and I saw a fellow coming down behind me hauling some kind of crude oil. It is pretty steep and he went right by me. I was doing maybe 12 miles an hour, and he was going maybe 60 miles an hour, brakes just smoking. He went right to the very very top, and I watched him and the thing broke loose — the tractor from the trailer. The trailer rolled all the way back down back to the highway and hit the highway with such a force that it just split the whole top of the tanker

wide open, black stuff just shot in the air all over and the driver ended up in the bush. I saw the same thing with a furniture van. There were fridges and stoves and beds going down the hill. Another time I was going down and a fellow with a load of hay was losing his brakes. He ran into the back of my tanker just to stop. Those days are all gone.

I guess the next challenge was to haul milk in Prince George, Vanderhoof, Smithers, a long way away. I have another story I've got to tell you. This was really something. George Johnson, the Dairy Inspector at that time, was in charge of all the north country including Smithers, Kitimat and so on. They had a tanker working out of Kitimat used to haul all the milk from Smithers. The road was just terrible, absolutely. You can't believe what the road was like. Twenty-three miles used to take us four-and-a-half hours. It's single lane, a thousand feet into the Skeena River, and up there against the mountains and so close to the ocean a lot of storms would come in. Well, some of them would come through the valleys and dump a foot of snow all at once, and then you go around the next corner and there would hardly be any. But the road was so bad, and the tanker used by the local dairy was splitting and breaking up. The milk would go down into the insulation and then when the truck would hit a bump the milk would come back out. It contaminated the rest of the milk in the tanker. The inspector really got on our case and said, "Look, you have to go out there and haul this milk." So anyway, I went up there and I was hauling the milk from Smithers to Kitimat. During one trip out to Kitimat, one of our drivers had an accident in

Unloading platform, Dairyland, 1963.

Pure Milk Dairy, Saanich

George Burgess started up a dairy farm about 1920 near the Oak Bay Boat House ... He was also working as a driver for the Victoria City Dairy ... When we spoke to him in 1963, Mr. Burgess told us of early one Good Friday morning during the time he was still a driver for Victoria City Dairy. He had a new horse on his two-horse team which kept trying to bite the other horse. While Burgess was making a delivery at St. Charles and Fort Street, in the narrow part of Fort Street known as the Dardanelles, something happened and the horses took off flying down Fort Street towards the Royal Jubilee Hospital corner of Fort and Richmond — George Burgess chasing after them. At the corner, the horses couldn't decide which way to go and managed to run the cart into a pole, spilling some 70 gallons of milk. A news boy found the second horse half an hour later. The next day, all the dairy's carts had either a brake or a hobble — but Burgess's cart had a brake and two fifty-pound hobbles.

Ron Greene, 2000.

Armstrong. It wasn't his fault at all. There was death involved. Of course his truck was seized. Now our spare tank truck was no longer available. So they phoned me and said, "Look, you've got to get back here." It's 4:00 in the afternoon when they called me, I'm at Kitimat, and we had just a terrible storm. It's on record as one of the worst they ever had. So all the rivers were flooding and the road was bad. I got back to Smithers about midnight and I was supposed to be back in Vernon by the next day and here I am stuck because now the road is closed. They had a Bailey bridge about 3,400 feet long over the river. I stopped my truck at the roadblock. A man said, "The water is right over the bridge and you can't get across." So I said, "Well, I'm going to go down to the bottom. There's a little bit of a campground there. I can go and get some sleep." "Yeah, go ahead," he said. So I went down there, and I looked at the river and it was just moving very slowly, debris flowing along, and I couldn't see the bridge because the water was just over the railings. So I took my pants off and my shoes and I walked across the bridge. The water was really cold. I walked across one side, and I turned around and the bridge was okay. There was no problem — it was just that you couldn't see it. So anyway, I thought I should be able to drive across it if I took the fan belt off and so on. Then I went up into that campground and I got a little stick with kind of a "y" on the end of it, and I cut it off so I could use it. Then I pulled partly onto the bridge and I got one end of that stick against the railing and the other end against the seat. I left my door open so that I knew exactly how far I

was from the railing. So I thought I'd just drive across very slowly. Well, it was a good plan alright, but the only thing I neglected was the doggone water was so deep that it came up into the floor mat of my truck and the floor mats started to float and interfered with my gas pedal. I couldn't work the gas pedal anymore. There I was trying to work the gas pedal with the mats floating, and trying to hold the stick. I couldn't use head lights or anything because the water was so deep. Anyway, I finally made it across, then I got to Houston. I got there at 3:00 in the morning to get fuel, and of course because the road was closed, the guy shut the station down. So I had to find out where he lived, go wake him up and tell him to go and get me some fuel. Then he wanted to know how I got there. He didn't believe me that I came across the bridge. So anyway, I got fuel, I kept going. I had been working for 30 hours, and I was so tired. I was coming into Prince George, and just as I came down the hill by the drive-in theatre there, the sun was coming up. I was absolutely gone. I had to pull in, so I pulled in to this drive-in yard, and shut the truck off, sleeping instantly. The wind was just tumbling, all part of this huge storm. Anyway, I didn't sleep very long. All of a sudden, there was a horrific bang, just like my cab had exploded and I jumped out of the seat, and of course my eyes didn't focus or anything and I don't even know where I was, and all of a sudden I realized the drive-in screen had blown over and just missed my truck by about 10 feet. It landed right in front of my truck. Boy, I was up and out of there. I was gone. I got back to Armstrong about 4:00 in the afternoon.

VICTORIA DAIRIES AGREEMENT

THIS AGREEMENT entered into this   day of   , 1962

BETWEEN:

NORTHWESTERN CREAMERIES LTD.,
1015 Yates Street, Victoria, B. C.

ISLAND FARMS DAIRIES CO-OPERATIVE ASSOCIATION,
2220 Blanshard Street, Victoria, B. C.

PALM DAIRIES LIMITED,
930 North Park Street, Victoria, B. C.

SHEPHERD'S DAIRY LTD.,
1645 Fort Street, Victoria, B. C.

(hereinafter called the "Employer")

OF THE FIRST PART

AND:

MILK SALES DRIVERS AND DAIRY EMPLOYEES' UNION,
LOCAL 464, of the Province of British Columbia,
affiliated with the International Brotherhood
of Teamsters, Chauffeurs, Warehousemen and
Helpers,

(hereinafter called the "Union")

OF THE SECOND PART

WITNESSETH THAT the Parties hereto agree as follows:

SCHEDULE "D"

RETAIL AND COMBINATION DRIVER SALESMEN - ROUTE FOREMEN

Classification	Wage Per Month October 1st, 1962
Retail and Combination Driver Salesmen	
Guaranteed Base Wage	$ 355.00
	207.00
Retail Units up to 7,000 per month	
Retail Units 7,001 to 9,000 per month	1½¢ per unit
Retail Units 9,001 and up per month	2¼¢ per unit
Cottage Cheese	2¼¢ per unit
	1¼¢ per pound or carton
Wholesale Units on Combination Routes	
Butter - 1¢ per pound - retail	1¢ per unit
½¢ per pound - wholesale	
Route Foremen	
Guaranteed Minimum wage or the average of all routes plus $45.00 whichever is the greater.	400.00

All Apartment House Routes - Additional $10.00 on basic wage.
In any district, a route with 50% or more of its customers in
apartments shall be classed as an Apartment Route.

SCHEDULE "E"

NON-COMMISSIONED DRIVERS

Classification	Wage Per Month October 1st, 1962
Truck Driver - Farm Pickup (with Certificate)	$
Hauling Truck Driver	355.00
Special Delivery Drivers	352.50
Maintenance Staff	
Maintenance Man	360.00
Engineers	401.50

Agreement between four Victoria dairies and the Milk Sales Drivers and Dairy Employees Union, 1962.

High Standards

In his 43 year career at Island Farms, Fred Mockford rose from one of the most junior positions to the most senior position — general manager. Along the way, he set the highest standards of performance, leading by example, and inspiring his fellow employees to follow suit.

Des Thompson, 2000.

Driving for Island Farms, 1940s

Fred Mockford, 2000.

The first job I had was driving a horse and wagon doing deliveries at fourteen years old. I wasn't old enough to have a drivers license! In those days, we used to start around four o'clock in the morning and, it was a long drive out to the end of the route, so we probably worked an average of ten or eleven hours a day. We used to work 42 days straight, seven days a week. Then we'd have a week off, providing there relief available. Horses used to have six days on, one day off. They took more days off than the driver! Routes were an average of 300 quarts a day. Since the average customer was probably an average of a quart a day, it meant serving about 300 customers. I believe I started at ninety-five dollars a month. We were responsible for our own credit. If customers had bad credit, we paid the price. In those days, we were delivering in big round bottles. Later on, they went to a flat sided square bottle for space efficiency, and then to cartons instead of bottles. The wagons were open. In winter time, in cold weather, we'd sometimes end up halfway through the day with the bottles frozen. If it was a real cold spell, we'd have a lot of cracked bottles. In summer, we would carry wet sacks to help cool the milk. Later on, we used ice.

The horses were stabled right in the downtown core on Cormorant Street, by City Hall in Victoria. We had three people on staff at the stables. There was always one man on in the morning and he and the driver hitched the horse up. We'd go from the stable to the dairy, and then load our own wagons. In those days,

there was a little cubby hole on Broughton Street downtown. The night shipper would wheel the milk out into the middle of the yard in different stacks, and we would pull up to our stack. There were probably a dozen horse routes on the road and very seldom a day went by that there wasn't a runaway with a horse. Such things as street car bells clanging would set them off. Most deliveries were at back doors. You would come out from behind the house, and no horse, no wagon. So you would have had to head back to the stable. That's where they would go.

I think most people were sad to see the horses go. The horse had become sort of a friend to customers. They'd bring treats out for the horse, which could be a problem, too. One time, I was in behind a house making a back door delivery and this lady always put a carrot out for the horse. Every morning. She was working out in the back yard when I went to make the delivery, and we got to talking. We heard a resounding crash and clatter and ran around to the front. The horse had gone through the gate, but his wagon hadn't.

The last horses were phased out after the war, in 1948, I believe. Later I was route foreman, distribution manager, assistant general manager, then eventually general manager. It was a great experience working for Island Farms. They were great people to work with.

From Hauling Cans to Early Tankers

Willoughby Rooke, 2000.

I started on the first of July in 1932. I was only around 19 then and I had an old Star Coupe. I traded that in for a GMC truck, one and a half to two ton. I went for

136

Mel Hand delivers milk in deep snow, 1951.

Moving Ice

Jim Calhoun started work for the FVMPA in 1930 and worked for them for the next 48 years. As a little boy, he lived close to the Chilliwack Creamery, located on the corner of Chilliwack Central and Young Street. When he was four, his mother would send him to the creamery with his wagon to pick up a block of ice which he would pull home for the ice chest. Archie Gould worked at the creamery and would sometimes help Jim pull his wagon up the hill leading from the creamery toward home.

Gerry Adams, 2000.

New tanker, Island Farms, driven by Cyril Shelford, Ministry of Agriculture, 1972.

43 years. I quit on the first of July 1975. I hauled independent milk. I did haul one FVMPA shipper, J.W. Berry, when he was producing Preferred Raw Milk. He bottled it and we picked it up in the morning. I took that into Dairyland, then I'd pick up the empties in late afternoon on the way out. I did haul his for quite a few years, then the Milk Board did away with Preferred Raw Milk.

To pick up cans, I had a flat deck, stake sides, drop tail gate. A can of milk weighed 127 pounds, full, and I was 125 pounds. It took me awhile to get on to it, but I could get it up onto the drop tail gate which was about half way. I eventually could lift one of them up on the full deck. I never did get any more than a 144 pounds. I was shipping to Melrose, Turner's, Crystal, Steves Dairy, Jersey Farms, Guernsey Breeders, Clark, Creamland Cresent. My route at the time was 40, 50 cans. In the spring, what they called the flush season, it would go up probably around 80 or 90 and then they would drop down again towards fall. I was finished about half past eleven. In those days, there was a lot of cord wood and stove wood cut around this area and I got hauling that to town. In the afternoon, I'd take two loads right into Vancouver. In 1955, I got my first tank. Then I bought out Sumas Transport and they had two tankers. I ended up with six tankers for a while.

We'd get a freeze up in the winter and, in this area of Langley, they only had one grader So we'd go as far as the chains would take us and then we'd have to dig the rest of the way to get in. We'd start out at 6:00 in the morning and 11:00 at night we'd come in and we still wouldn't have a load. It all depended on the wind. If you get the wind, you could head up these roads and they'd be almost bare. The next mile you'd have to get out and dig. Sumas was the worst. You could stand on top of drifts and put your hand on the transformers. I remember going into one place and all I could see was the shape of the door. He was completely covered. We had to have snow plows and bulldozers to bust through that stuff. Vedder Trucking had a tandem dump truck and they had a snow plow on the front of it. That's all they kept it for, the winter time. They might need it for two or three days or they might need it for two weeks, they never knew.

I started with a 1500 gallon tank. I bought a little International with the motor underneath and it was just a short thing. It was the greatest little truck for getting around. I did that till it wasn't big enough anymore and I had to get a bigger one. When I had the little truck, the tanks were about two inches thick — they were all cork between two tanks. On the hottest day of summer, we finished picking up just around noon and it sat out beside the house till the next morning. At 5:00 in the morning I'd head to town with it, and it wouldn't go up any more than about one or two degrees, that's all. ■

Truck Fleet, Northwestern Creamery, Victoria, 1946. Photo by Duncan MacPhail.

Drivers, Northwestern Creamery, Victoria, 1946. Photo by Duncan MacPhail.

CHAPTER V

The Clyne Commission

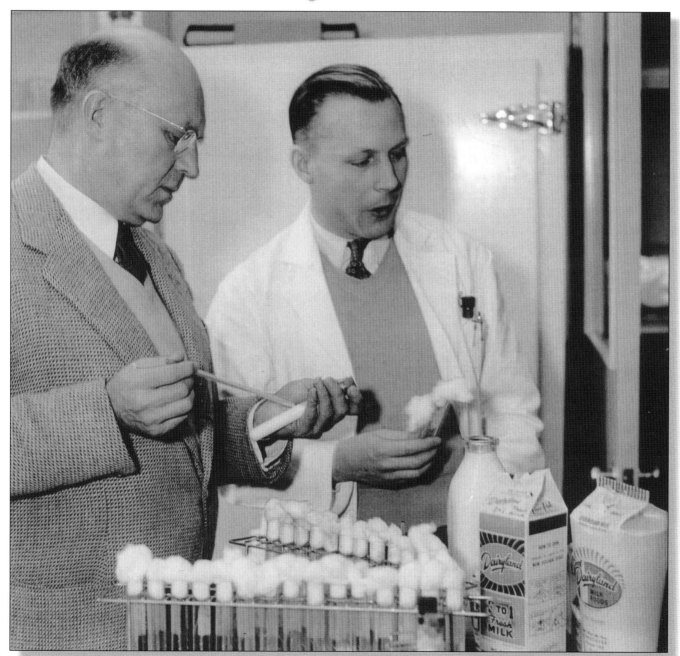

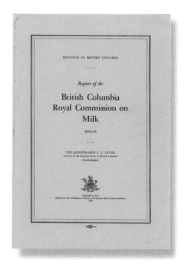

Neil Gray, Dairyland bacteriologist, chats with Justice Clyne, 1955.
BC FARM MACHINERY MUSEUM COLLECTION

Cover, report of the Clyne Commission, 1955.

I believe that a new Act is necessary ... I think that it is clear from the evidence which I have heard that British Columbia is lagging behind other parts of this continent in modern concepts of milk-marketing, and I believe that industry is due for a general house-cleaning, which may be accomplished by new legislation such as I have suggested. I have taken the liberty of expressing my views somewhat freely upon proposed changes, but I would not have done so if I had not had the opportunity of forming my opinions from the testimony of a great number of very competent and intelligent witnesses who gave their evidence clearly and frankly, and to whom I express my gratitude.

J.V. Clyne,
Royal Commission
on Milk, 1956.

On September 3, 1954, the Honourable J.V. Clyne, a judge of the Supreme Court of British Columbia, was asked by the Province of British Columbia to oversee a royal commission designed to look into what seemed like a simple question. As Commissioner, Clyne was asked to find out whether milk producers in the Vancouver area were receiving the price fixed for their milk by the Milk Board under the Public Utilities Act. This order required distributors to pay producers of fluid milk a minimum of $5.03 per hundredweight on the basis of 3.5 percent butterfat content and to adjust this price with a premium or a discount for a greater or lesser percentage of butterfat. For industrial milk — milk sold to be manufactured into products such as butter, ice cream, cottage cheese, evaporated milk and milk powder — a minimum of $1.96 per hundredweight had been fixed. The commission was an acknowledgement by the provincial government of the political importance of milk. Sixty percent of the population of the province lived in the Fraser Valley milkshed; thus, the province was trying to do many things. It wanted to foster the growth of the dairy industry, which required a stable, profitable marketplace for producers and distributors and a reasonable price for consumers. It wanted to create a careful balance in the industry somewhere between an

Monthly report, BC Lower Mainland Dairy Products Board, 1936.
ACTON KILBY PAPERS, KILBY FARM MUSEUM

adequate and an excessive supply of milk. It also wanted to build in breathing room to allow the industry to expand as population grew.

The answer to the pricing question that framed the royal commission was not simple, and the recommendations Clyne returned had weighty consequences for the dairy industry in all parts of British Columbia. The wrangling between distributors and producers that precipitated the striking of the Clyne commission was based on perceived rights and responsibilities with regard to markets and surplus milk in the immediate area of Vancouver. But the grounds on which these milk wars had been waged were an undefined territory between federal and provincial jurisdictions. The constant

Vancouver, B.C., November 23, 1938.

To All Independent Dairy Farmers and Producer-Vendors on the B.C. Lower Mainland

SINGLE AGENCY

VOTE FOR BASIL GARDOM

(He was nominated by the Independent Milk Producers Co-Operative Association and other Independents.)

One Independent only can be elected to the single agency and this vote is being taken before the vote for the Independent member of the Milk Board, for which Mr. Sam H. Shannon has been nominated.

WE SOLICIT YOUR VOTE (YES "X") FOR BASIL GARDOM BECAUSE:---

1. HE HAS NO INTEREST IN ANY DAIRY OR FACTORY.

2. HE IS A PRODUCER AND WORKS FOR THE PRODUCERS, FIRST, LAST AND ALL THE TIME.

3. HE HAS THE ABILITY AND EXPERIENCE TO SECURE A SQUARE DEAL FOR ALL PRODUCERS AND PRODUCER-VENDORS.

T. H. McDONALD, Chilliwack
M. B. McDERMID, Eburne
A. E. J. FARROW, Chilliwack
W. M. OLDFIELD, Twigg Island
C. H. EVANS, Sardis
A. ALDRIDGE, Coquitlam

G. H. MILLER, Lulu Island
G. W. SHANNON, Cloverdale
WM. MONTGOMERY, Ladner
JOHN OLSEN, Matsqui
G. WORTH, Dewdney

Promotional handbill, sent to independent shippers, regarding the nomination of a representative to the Milk Board, 1938. ACTON KILBY PAPERS, KILBY FARM MUSEUM

Cover page, the *Industry*, published by Basil Gardom of the Independent Milk Producers Cooperative Association for the benefit of independent shippers, 1937. GARDE GARDOM COLLECTION

GUERNSEY
HOLSTEIN
AYRSHIRE
JERSEY

INDUSTRY

Devoted to the Interests of
THE PRIMARY PRODUCERS OF WESTERN CANADA

"Here shall the press the Peoples right maintain, Unawed by influence and unbribed by gain; Here Patriot Truth her glorious precepts draw, Pledged to Religion, Liberty and Law." - - Joseph Storey A.D. 1796

VOL. III.　　　　OCTOBER 15, 1937　　　　No. 7

INDIAN SUMMER on the B.C. COAST

Primary Industries Supply All Necessities of Life

I was called to Carr's office this morning for a discussion. There was nothing satisfactory. Everything is "if" and "and." We had a long discussion on the Quota Committee. I see nothing but trouble for me under the proposal as set out by Carr in which they want me as representing the Agency. I do not wish to be on the Board as it is to be set up. I wish to be left out of all Quota Committees and I have told Carr this. The whole situation is not good and no one with any power is trying to do anything to clear it up.

Acton Kilby, Notes to Directors of the Milk Shippers Agency, January 11, 1954.

testing and litigation between the independent producers and the cooperative members throughout the first decades of the twentieth century were important parts of the process of defining legal territory that had never before been defined in Canada.

John Valentine Clyne was born in 1902 in Vancouver, graduated from the University of British Columbia, and was called to the bar in 1927. He practised law for 20 years before being appointed Chairman of the Canadian Maritime Commission in 1947, a position he held for a few years before returning to the legal sphere to become a judge of the Supreme Court. His *Royal Commission on Milk* was not the first such task he had been asked to undertake. In 1953, he was commissioner of the *Inquiry into the Circumstances of Landslides at Whatshan*, a commission struck after two landslides destroyed part of a BC Power Commission hydroelectric plant at Whatshan in the Kootenays.

For the *Royal Commission on Milk*, Clyne was given a question and then given a great deal of latitude to find the answers. In short, he was allowed to "make such … enquiries as seemed to be desirable in the public interest"[1] and to "report thereon as speedily as possible to the Lieutenant-Governor in Council with the opinions which he may have formed … and with such recommendations as he may think proper."[2] His work was in part historical. He needed to sketch out the historical problem that saw the same issues continuing to surface over the years. He was empowered to call witnesses and, in fact, saw 143 of them during hearings totalling 71 days — 62 days at the court house in Vancouver, two days in Victoria, five days in Chilliwack, and one day

each in Ladner and Mission — all the while amassing transcripts of evidence totalling 10,565 pages. He accepted 55 briefs on behalf of a range of individuals and organizations, from producers and processors to consumer advocacy groups like the provincial branch of the Canadian Association of Consumers, and the Vancouver Local Council of Women. Clyne also took his questions on the road, travelling to plants, dairies, and farms in the Fraser Valley and mulling over the intricacies of the industry for more than a year before tendering his final report.

In his initial response to the question posed to him, Clyne found that for a number of years "there had been more or less constant disobedience to Board orders."[3] He soon realized he would have to study the industry in detail in order to find out why violations of Milk Board rulings were rampant in the industry, especially in the Fraser Valley. His task then, as he saw it, was to provide a comprehensive analysis of the dairy industry, with particular reference to "The City of Vancouver, the City of North Vancouver, the District of North Vancouver, the Municipality of West Vancouver, the Municipality of Burnaby, the Municipality of Richmond, the City of New Westminster, the University area in Point Grey, and that portion of the Fraser Valley which lies south of the 49° 30′ parallel, in the Province of British Columbia."[4]

By the time of the commission, the basic players in the industry in the Fraser Valley were the FVMPA; the Milk Shippers Agency, headed by Acton Kilby and comprising the distributors Avalon Dairy, Creamland Crescent Dairy, Jersey Farms, and Palm Dairies; and the

CLASS OF SERVICE DESIRED
Full-Rate Message
Day Letter
Night Message
Night Letter

Patrons should mark an X opposite the class of service desired; OTHERWISE THE MESSAGE WILL BE TRANSMITTED AS A FULL-RATE TELEGRAM.

FORM 6102

CANADIAN NATIONAL TELEGRAPHS

D. E. Galloway, Assistant Vice-President, Toronto, Ont.

Exclusive Connection with WESTERN UNION TELEGRAPH CO.
Cable Service to all the World
Money Transferred by Telegraph

RECEIVER'S No.	TIME FILED	CHECK

Send the following message, subject to the terms on back hereof, which are hereby agreed to

December 19, 1941

A.D. Paterson
Ladner, B.C.

WARTIME PRICE BOARD ANNOUNCES TONIGHT SUBSIDY TO MILK PRODUCERS OF THIRTY CENTS PER HUNDRED POUNDS ABOVE PRESENT LAWFUL PRICE FOR FLUID MILK IN AREAS WHERE NO INCREASE IN PRICE SINCE AUGUST FIRST THIS YEAR AND FORTY CENTS ABOVE MINIMUM PRICE OF ONE DOLLAR SEVENTY CENTS PER HUNDRED POUNDS ON MILK SOLD TO MANUFACTURERS STOP SUBSIDY WILL BE ADDED TO PRICE PAID BY DISTRIBUTOR OR MANUFACTURER WHO WILL BE REIMBURSED MONTHLY STOP PLEASE TELL DAVIE STOP REGARDS

IAN MACKENZIE

as from Dec. 22/41

Wartime Prices and Trade Board announces change in pricing, 1941.
ACTON KILBY PAPERS, KILBY FARM MUSEUM.

Advertisement, the Wartime Prices and Trade Board, *Alaska Highway News,* **14 September 1944.**
NORTH PEACE HISTORICAL SOCIETY

...about the things you buy in wartime

Milk IS NOT ELASTIC

PRODUCE for VICTORY

The production of milk and milk products in Canada has achieved a notable record. Careful planning makes the best use of every quart. But we must all recognize the conditions imposed by war and realize that in these times we all can't get all we want.

Oberson
Chairman
Wartime Prices & Trade Board

Milk is our most valuable food. It must provide us with milk to drink, with butter, cheese, evaporated and concentrated milk and a score of other products.

Canadian farmers have done a magnificent job. They have increased milk production in 1943 by more than 540 million quarts over 1938.

They have done this in the face of a 25% reduction in farm help, equipment shortages, and the fact that it takes two to three years to bring a calf into milk production.

WHAT HAPPENS TO ALL THIS MILK?

In spite of greater production, the demand for milk and milk products has risen even more, because—

(a) There's more money to spend;

(b) More people are working, with changed food habits and increased food needs;

(c) Our Armed Forces and Allies make heavy demands.

It has therefore been necessary by rationing, by subsidies, by careful planning and by other controls—to divert our milk supply into channels most suited to our various food needs.

The above graph shows in percentage how the total Canadian milk supply is used.

MISCELLANEOUS MILK 6% — EVAP. MILK 10% — CHEESE — FLUID MILK 35% — BUTTER 48%

FLUID MILK
TAKES 35% OF OUR MILK

Because fluid milk is regarded by nutritionists as the most nearly perfect food, nothing has been allowed to interfere with its sale. Today, Canadians are drinking more milk and a greater percentage of our milk supply is being consumed as fluid milk—than ever before. Fluid milk has the right of way, but don't waste a drop of it.

IT TAKES 9 QUARTS OF MILK TO MAKE ONE POUND OF BUTTER

CHEESE
TAKES 10% OF OUR MILK

CANADA'S CHEESE PRODUCTION
1943 164,500,000 LBS.
1938 127,600,000 LBS.

Canada's annual cheese production has gone up by about 37 million pounds since the war.

Cheese is a concentrated food product—easily shipped and stored. That is one reason why we send large quantities to Great Britain to help meet its pressing food needs. While in the last year we exported four of every five pounds of cheese we made, our production has been so increased that we have left for domestic use about three million pounds a year more than before the war.

BUTTER
TAKES 48% OF OUR MILK

In the first three years of war, our butter consumption increased 10.9%. So, rationing was established to prevent too much milk going into butter, at the expense of other important milk products, and to insure a fair share to everyone.

The rationing of butter was influenced by the fact that it has less nutritive value than some other milk products, and because we get a generous supply of fats or their food equivalent in other forms.

To maintain a proper balance of consumption between various milk products and to ensure that butter is put into storage for winter use—when production drops—it is necessary to reduce the ration from time to time.

The only Canadian-made butter not consumed in Canada is that sent by the Red Cross in prisoner-of-war parcels, each of which contains a 1 lb. tin.

EVAPORATED MILK
TAKES 6% OF OUR MILK

More than twice as much (152 million lbs.) was used by Canadians this last year as in 1938 (74 million lbs.). And yet, there has not been enough.

Where has it all gone? It's an important food for babies—and there are 50,000 more of them a year. Larger quantities have gone to areas where fresh milk is not available. Armed Services have added new demands. In spite of this, our exports of evaporated milk to Great Britain, Newfoundland and the West Indies, etc., are maintained at pre-war level.

ICE CREAM AND OTHER MILK PRODUCTS
TAKE ABOUT 1½% OF OUR MILK

The manufacture of ice cream is restricted to the 1941 level because milk is needed for other purposes.

Milk Powder and Condensed Milk are taking more milk. Milk sugar is used in the manufacture of Penicillin and for other wartime purposes. Casein (a milk by-product) is used in making glue for aeroplanes.

90% OF THE MILK PRODUCED IN CANADA IS CONSUMED IN CANADA

THE WARTIME PRICES AND TRADE BOARD

THIS IS THE FIRST OF A SERIES OF ADVERTISEMENTS GIVING THE FACTS ABOUT THE SUPPLY SITUATION OF VARIOUS WIDELY USED COMMODITIES

The Turning Point

The Clyne Commission was the turning point for our industry. The Clyne report made it possible for all dairy farmers to be treated fairly within this province regardless of to whom they shipped their milk. Up until that point, 1956, we all know the story of the dairies in the Vancouver area receiving their milk from farms on Lulu Island, from Richmond and Ladner — areas close to the city. In those days milk production at farms in the summer months was about double the production in the winter months. The high point of production was usually about the end of May and the low point of production at farms was about the middle of December. If there were a million pounds produced on a day in June, there would be 500,000 pounds produced in a day in December — it was that serious. The dairy processors within the Vancouver area would have an adequate supply from their own producers during the summer, but would be required to draw supplies of milk from the Fraser Valley Milk Producers' Association for the balance of the year. The Fraser Valley members were receiving milk from all areas of the Fraser Valley and were providing the manufacturing plants at Sardis and Delair with large volumes of summer milk to process into skim milk powder, butter, and Pacific evaporated milk. They were, in fact, taking care of all the milk surplus to the needs of the fluid market and were called upon to supplement the fluid needs of the Vancouver dealers during those periods of short supply. Thus, only a small portion of the Fraser Valley members' milk was entering the higher value fluid market, when compared to the independent producers who received the fluid milk return for their entire production.

The Clyne Commission recognized that all dairy farmers should be treated equally and fairly, dependent upon them providing the proper facilities on their farm. The report recognized that if you were going to be treated fairly and were going to derive your income in a pooled form from the marketplace, then you had to produce milk of a sufficiently high standard to meet the needs of the fluid market. A variety of programs were introduced. Regulations relating to the premises at the farm, the type of building, the hot water system, the temperature of milk at farm and upon arrival at the processing plant, and the bacteria count were all considered and recommended, as were proper provincial government inspection and reporting procedures. In other words, all licensed producers would be required to meet these new standards regardless of whether their milk entered the fluid market or was processed into powder, butter, cottage cheese, or evaporated.

People like Ken Savage and Herb Rhiel played significant roles. Ken Savage was the Provincial dairy commissioner at the time and Herb Rhiel was his assistant. They were both actively involved in developing the standards now in place within our industry, deriving much of their information from systems in place in the United States.

Neil Gray, 2000.

Independent Milk Shippers Cooperative Association, headed by Basil Gardom. The roots of their disagreements dated back to the time of the construction of the BC Electric Railway from its terminus in Chilliwack through the communities of the Fraser Valley and into Vancouver, which enabled the shipment of milk from Chilliwack in 1910. Suddenly, Vancouver milk dealers had a lever in the form of this new source of supply to force farmers in Ladner and Delta — the westernmost portion of the Vancouver milkshed — to lower their prices. The formation of the FVMPA in the era of the First World War held out the hope of cooperative bargaining power for many farmers. However, even by 1917, when farmers representing 90 percent of the milk production in the Fraser Valley area had joined the Association, it remained unable to dominate the supply to Vancouver milk dealers and to set what it saw as fair prices to producers. Vancouver dealers fought to keep the FVMPA from monopolizing the supply market to Vancouver by continuing to buy from a range of producers of their own choice and by offering premiums to producers who would leave the umbrella of the FVMPA. At the same time, the FVMPA got aggressive on the distribution front by buying a number of Vancouver milk distributors and creating Fraser Valley Dairy Limited, a subsidiary of the FVMPA that competed vigorously with remaining independent dealers. The FVMPA further strengthened its position by buying the Pacific Milk Company and by building a general utility plant at Sardis. This plant was a benefit to all milk producers — independents and cooperative members alike — because it absorbed surplus milk, allowing

Phone: 12 Noon
August 4/52

C.P.Telegraph   VICTORIA

ACTON KILBY

RE TEL YOU MAY BE ASSURED THAT YOU WILL BE GIVEN OPPORTUNITY

TO DISCUSS ANY PROPOSED MILK BOARD CHANGES WITH ME BEFORE

SUCH CHANGES GO INTO EFFECT.

W. K. KIERNAN
MINISTER OF AGRICULTURE

Telegraph from Minister of Agriculture Ken Kiernan to Acton Kilby, 1952.   ACTON KILBY PAPERS, KILBY FARM MUSEUM

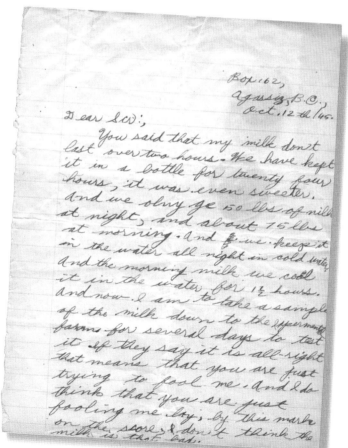

Letter to Acton Kilby, Milk Shippers Agency, protesting the agency's quality control measures, 1945.

ACTON KILBY PAPERS, KILBY FARM MUSEUM

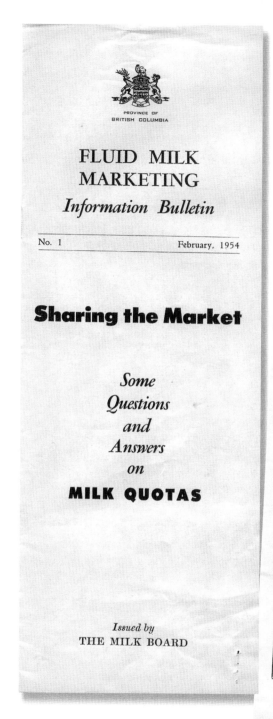

PROVINCE OF BRITISH COLUMBIA

FLUID MILK MARKETING
Information Bulletin

No. 1   February, 1954

Sharing the Market

Some Questions and Answers on

MILK QUOTAS

Issued by
THE MILK BOARD

Foreword

As Minister of Agriculture, I welcome Milk Board Information Bulletin No. 1. Its contents, I am sure, will be well received by those engaged in the fluid milk industry, and by the general public.

Since the promulgation of Order No. 40 of the Milk Board, many questions have been asked respecting certain aspects of the Order.

These questions have come from many sources and have been diverse in character. It has been noticed, however, that the greater number have dealt with those provisions setting up Quota Regulations, and it has become apparent that misunderstanding has developed.

In order that producers and others may have a better knowledge of the intent of these provisions, the Milk Board has prepared this bulletin. It answers questions most commonly asked.

It is realized that the subject is not entirely covered herein. However, as the occasion arises, similar bulletins will be issued to deal with any matter relative to the fluid milk industry which is of current interest. In the meantime you are invited to submit further questions or suggestions regarding matters you would like dealt with to the Milk Board. Do not depend on hearsay or rumour for your information.

It must be recognized that any Act, Regulation, or Order is properly subject to review and change in the light of necessity as dictated by actual experience. Change, however, must only be made when it becomes apparent that the interests of the majority affected are best served by such change.

KENNETH W. KIERNAN,
Minister of Agriculture.

Information bulletin issued by the Milk Board, six months prior to the beginning of the Clyne Commission, 1954.

GEORGE AYLARD COLLECTION

147

Preserving Our Identity

The Independent Milk Producers' Association, comprising a small number of dairy farmers, obtained their charter in 1930, and it has been my privilege to act as their president from August, 1931, to this date. At first we marketed our product individually to independent dairies which required quality milk for the fluid trade; since February 1, 1935, we have wholesaled as a voluntary cooperative group to the seven independent dairies which, with us, have fought to preserve our and their identities.

Basil Gardom, The Industry, February 1938.

Political cartoon, late 1940s or early 1950s, by Ernie Walker.

the fluid price to be kept higher than the manufactured price.

Because independent producers sold only to distributors serving the fluid market, they remained able to sell their milk at a much higher price than members of the FVMPA. This situation, according to Clyne in 1956 "has created between the two producer groups an intense bitterness which has existed up to the present time."[5] The relationship between the two groups was complex. Members of the FVMPA collectively had made a commitment to managing the surplus milk of

their membership, and now found themselves also managing the surplus of the independent shippers — buying surplus in months of plenty, and selling milk when the Independents had a shortage.

The first twinges of legal trouble came in the 1920s: in 1927, the *Produce Marketing Act* was created "to deal with the sale of all agricultural production and to apply where a large majority of producers favoured its implementation."[6] Milk products, however, were not included under this act. Trying to provide protection for farmers, E.D. Barrow, the Minister of Agriculture who

IN THE MATTER OF THE "PUBLIC UTILITIES ACT"

and

IN THE MATTER OF THE "PUBLIC INQUIRIES ACT"

and

IN THE MATTER OF A COMMISSION HELD TO ENQUIRE
INTO CERTAIN PHASES OF THE MILK INDUSTRY IN
BRITISH COLUMBIA

- - - - - - - - - -

TO THE HONOURABLE MR. JUSTICE CLYNE

Sir:

The "Milk Shippers Agency" submit the following observations. In view of the mass of material submitted, we will attempt to keep our comments and suggestions to the point.

We are totally against one "Big Pool for Milk".

We believe that such an arrangement would create a monopoly, which in the hands of unwise guidance, could do untold harm to the Producers and be difficult to rectify.

We believe the present system of a Producer shipper having the right to choose his own marketing agency is a fundamental right.

The Price of Fluid Milk should be set by the Milk Board and should apply right through from the Producer to Consumer.

The much discussed "Store Milk" Sales are only a partial answer.

It must be realized that elderly people and mothers with small children cannot leave the home to go to a store;

in the case of children, the risk of fire is apparent.

The Dealers are entitled to a just spread for the functions performed by them and such spread should be set by the Milk Board.

The supervising of the Grading, Weighing and Testing of Milk should be the responsibility of the Provincial Government, Department of Agriculture, Dairy Branch.

We do not believe it is necessary to "Bond" Milk Dealers. The simple equation of no pay, no milk, will readily apply.

We recommend that a representative of the B. C. Federation of Agriculture be appointed as a member of the "Milk Board". This organization represents the producers.

In considering these problems, it should be remembered that the Producers are entitled to equal sympathy as that asked for by the Consumers and employees.

In conclusion we would point out that the Milk Shippers Agency consists of 242 Milk Producing Farmers. They appoint their officers and set their policies at an Annual Meeting and have done so since 1935.

DATED this 21st day of January A.D. 1955.

Respectfully submitted,
MILK SHIPPERS AGENCY.

Submission to Clyne Commission by the Milk Shippers Agency, 1955.

Less Government Control

What British Columbia needs today is less talking and more doing, fewer exponents of theory, less government control, less restriction of trade and more freedom for labour and capital in primary industry which creates all new wealth.

The principle of voluntary marketing:

- It is freedom of choice of practical plan.
- It is orderly and lawful promotion.
- It is strict regulation of quality and service.
- It is the maintenance of bargaining power.
- It is a fair price to the producer.
- It is a reasonable price to the consumer.
- It is the opening of all channels of trade.

Energetic advancement of this policy will soon bring back work and wages, restore the morale of all true Canadians and reduce the government's debt.

Basil Gardom, "Compulsion in Primary Industry is Loss of Free Citizenship," a radio address over CBR, November 26, 1937.

had brought forward the bill, created another bill called the *Dairy Farmers' Losses Redistribution Bill.* Reaction to this bill resonated through the dairying community as well as through the halls of the legislature. A royal commission was struck and headed by Dr. Clement of UBC. It recommended that "the production, distribution, and sale of milk and cream for the fluid trade be treated as a public utility, to be closely regulated and safeguarded in the interests of the public as a whole."[7] His rationale was that farmers were entitled to "the same rights and privileges as other business organizers of the country," and that regulated marketing and legislation to ensure equalized access to the fluid market would put him in "a position to render a large service to himself and the consuming public."[8]

Before Clement's report became public, however, the British Columbia Court of Appeal upheld the legality of the original *Produce Marketing Act.* The FVMPA was pleased. A few months later, in March of 1929, J.W. Berry brought forward a private member's bill, called the *Dairy Sales Adjustment Bill,* designed to draw milk producers under the protection offered by the *Produce Marketing Act.* It passed, but the act could only come into effect under stringent conditions that required the FVMPA to work with Independents to create stability in the marketplace. "Soon after the legislation was passed, the FVMPA directors met with a group of independent distributors … to negotiate an agreement. Meetings were held throughout the Valley. An overwhelming majority of cooperative members and a majority of the Independents, reorganized as the Independent Milk Shippers' Association, voted in

favour of petitioning the government to immediately implement the *Dairy Products Sales Adjustment Act* immediately. The FVMPA elected A.H. Mercer as their representative on the Milk Board set up under the provisions of the Act. The Independents chose Samuel H. Shannon. Charles A. Welsh of New Westminister, a choice acceptable to both groups, was appointed chairman. The nature of the Milk Board with its representative function reflected the assumption that the complexities of milk marketing could be handled by a process of negotiation between two groups of producers."[9]

On January 1, 1930, milk control came into effect. For Independents, the pooled price represented a significant drop in income; for FVMPA members, the reverse was true. Many producer-vendors simply opted out by claiming they were selling preferred raw milk — an area not regulated under the new system of levies. Secure in the validity of the legislation, Milk Board members decided to prosecute four farmers who refused to comply. When their case was thrown out on a technicality, the Board's inability "to enforce the provisions of the Act was an ominous warning that resistance to the legislation would eventually render it inoperative."[10] According to historian Morag Maclachlan, decisions made in the milk industry were related to those made in the tree fruit industry, largely concentrated in the Okanagan. When a grower, A.O. Lawson, challenged the *Produce Marketing Act* on the grounds that it was a form of indirect taxation to the farmer, the provincial judiciary upheld the act. In 1931, when he took his case to the Supreme Court of Canada, the Supreme Court

1. Aldersey, Paul, Sunny Brae Dairy Limited, Duncan.
2. Anderson, Aage, Dairy-farmer, Deroche.
3. Anderson, Miss Gladys D., Accountant, Purdy's Café, Vancouver.
4. Anderson, Dr. Walton J., Associate Professor and Chairman, Department of Agricultural Economics, University of British Columbia.
5. Armishaw, Herbert, Producer-Vendor, Nanaimo.
6. Armstead, Daniel M., President, Seal-Kap Dairy Limited.
7. Atkinson, Lyle A., Assistant General Manager, F.V.M.P.A., and Manager, Dairyland Division of F.V.M.P.A.
8. Barker, Archibald S., Dairy-farmer, Chilliwack.
9. Barnes, Robert J., Dairy-farmer, Pitt Meadows, and President, Farmers' Institute, Pitt Meadows.
10. Bartlett, Dr. Roland W., Professor of Agricultural Economics, University of Illinois, Chicago.
11. Bell, Gordon T., Inspector, B.C. Milk Board.
12. Blair, David J., Dairy-farmer, Westham Island.
13. Block, Jesse F., Dairy-farmer, Port Kells District, and Member of Milk Committee, District " E," Farmers' Institute.
14. Blumes, Dr. Joseph, Consumer, Vancouver.
15. Borradaile, Osmond H., Dairy-farmer, Chilliwack.
16. Brown, David, Dairy-farmer, Sumas Prairie.
17. Brown, William G., Sales Manager, Peters Ice Cream Company Limited.
18. Bryson, Dr. Henry L., Chief Food and Dairy Inspector, City of Vancouver.
19. Burdge, Francis W., Dairy-farmer, Victoria.
20. Burwash, Thomas A., Manager, Jersey Farms Limited.
21. Campbell, Dr. John J. R., Professor of Dairy Bacteriology, University of British Columbia.
22. Carr, Ernest C., Chairman, B.C. Milk Board.
23. Carradice, Francis W., Consumer, Vancouver.
24. Chester, Dr. Kenneth, Veterinary Surgeon, and Registrar, B.C. Veterinary Association.
25. Clarke, Dr. Mills F., Superintendent, Dominion Government Experimental Farm, Agassiz.
26. Clark, Dr. Robert M., Associate Professor of Economics, University of British Columbia.
27. Clement, Dean (Emeritus) F. M., Consultant.
28. Crawford, William, Producer-Vendor, Cloverdale.
29. Crowley, Everett, Manager, Avalon Dairy Limited.
30. Currie, Daniel, Dairy-farmer, Lulu Island.
31. Daum, Elmer D., Accountant, B.C. Milk Board.
32. Davie, Murray A., Dairy-farmer, Ladner.
33. Davis, Leo, Dairy-farmer, Mission.
34. Dewar, James E., Assistant Business Agent, Milk Drivers and Dairy Employees' Union, Local 464 of the International Brotherhood of Teamsters, Chauffeurs, Warehousemen and Helpers.
35. Dissing, Svend, Department Manager, Acme Dairy Limited.
36. Dolman, Dr. Claude E., Professor and Head of Department of Bacteriology and Immunology, University of British Columbia.
37. Drake, Wilson B., Vice-President, Drakes' Dairy Limited.
38. Eaton, Mrs. Fraudena, Member, B.C. Milk Board.
39. Edgar, Alexander W., Secretary, Vancouver Milk Distributors' Association.
40. Edwards, Thomas, Director, F.V.M.P.A.
41. Elgood, Mrs. Mary M., Dairy-farmer, Abbotsford.
42. Fawcett, George, Manager, Palm Dairies Limited.
43. Flowerdew, Eric S., Dairy-farmer, Aldergrove, and Member of Special Milk Committee, District " E," Farmers' Institute.
44. Forsberg, Walter K., Purchasing Agent, Vancouver General Hospital.
45. Friesen, John V., Dairy-farmer, Ladner.
46. Gardom, Basil, President, Independent Milk Producers' Co-operative Association.
47. Gardom, Basil, Speaking for Acme Dairy Limited.
48. Gartshore, Alexander C. W., Restaurant Proprietor, Vancouver.
49. German, Harold W., Dairy-farmer, Chilliwack.
50. Gilbert, James L., Dairy-farmer, Aldergrove.
51. Gibbs, Charles D., Dairy-farmer, Sumas Prairie.
52. Goepel, Moffat P., Comptroller, F.V.M.P.A.
53. Grant, Neil H., Sales Manager, Jersey Farms Limited.
54. Grant, Ronald D., Regional Supervisor, Veterans' Land Act, New Westminster, B.C.
55. Grauer, John J., General Manager, Frasea Farms Limited.
56. Grimston, Douglas G., Manager, Valley Ice Cream Limited.
57. Gunn, Dr. Wallace R., B.C. Live Stock Commissioner and Chief Veterinary Inspector.
58. Haslam, Charles A., President and Managing Director, Glenburn Dairy Limited.
59. Hay, Kenneth, Producer-Vendor, Sunnybrook Farm Dairy, Vancouver.
60. Headey, Cyril, Dairy-farmer, Port Kells, and President, Port Kells Farmers' Institute.
61. Helliwell, John L., C.A., Accountant to Commission.
62. Henderson, Allan O., Dairy-farmer, Ladner.
63. Hogan, Alexander, Dairy-farmer, Sumas, and Reeve of Sumas.
64. Homan, Henry J., Dairy-farmer, Lulu Island.
65. Honeyman, John D., Dairy-farmer, Ladner.
66. Honeyman, Stuart N., Dairy-farmer, Ladner.
67. Hoopes, Lorenzo N., Plant Operations Manager, Lucerne Milk Company (United States and Canada).
68. Hoy, Norman D., Co-Manager, Royal City Dairies Limited.
69. Hulbert, John, Dairy-farmer, Koksilah, and Representing Cowichan Agricultural Society.
70. Hurford, Richard U., Dairy-farmer, Courtenay, and Secretary, Comox Co-operative Creamery Association.

71. Kadla, Frank J., Student, Agricultural Economics, University of British Columbia.
72. Kidd, Dr. Abraham, Assistant Live Stock Commissioner and Assistant Chief Veterinary Inspector.
73. Kilby, Acton, President and Managing Director, Milk Shippers' Agency.
74. King, Professor (Emeritus) Harry M., Member, B.C. Milk Board.
75. Kournossoff, Michael V., Dairy-farmer, Chilliwack.
76. Kraft, Walter J., Vice-President and Division Manager, Canada Safeway Limited.
77. Lambrick, Arthur G., Producer-Vendor, Victoria.
78. Langford, Carl W., Dairy-farmer, Aldergrove.
79. LeWarne, Alfred, Manager, Sno Freze Ice Cream Company Limited.
80. Lillico, Mrs. Lydia V., Past Provincial President and Director on National Board, Canadian Association of Consumers.
81. Macken, William L., Former Dairy-farmer, Chilliwack, and Former Manager, F.V.M.P.A.
82. Mather, Dr. James M., Professor and Head of the Department of Public Health, Faculty of Medicine, University of British Columbia.
83. Mercer, A. H., General Manager, F.V.M.P.A.
84. Mills, Mrs. Elizabeth, President, Victoria Branch, Canadian Association of Consumers.
85. Moynes, James M., C.A., Helliwell, Maclachlan & Company.
86. Murray, Alexander B., Business Agent, Milk Drivers' and Dairy Employees' Union, Local 464 of the International Brotherhood of Teamsters, Chauffeurs, Warehousemen and Helpers.
87. Murray, Dr. Sydney S., Medical Health Officer, City of Vancouver.
88. Mutter, James A., Manager, Cowichan Creamery Association.
89. McArthur, Herbert E., President and Manager, Creamland Crescent Dairy Limited.
90. MacAulay, John A., Q.C., Vice-President and Director, Canada Safeway Limited.
91. McCreary, Dr. John F., Professor of Pædiatrics, University of British Columbia, and Head of Health Centre for Children, Vancouver General Hospital.
92. McDermid, Matthew B., Dairy-farmer, Lulu Island.
93. MacDonnell, Geoffrey, Dairy-farmer, Chilliwack.
94. McKinnon, Norman L., Producer-Vendor, Port Alberni.
95. McMynn, Ewart, General Manager, Richmond Milk Producers' Co-operative Association.
96. Nelson, Charles E., Former Manager, Guernsey Breeders Dairy Limited.
97. Nicholson, Daniel R., President, F.V.M.P.A.
98. Norman, Leonard J., Dairy-farmer, Sumas Prairie.
99. Okulitch, George J., Production Manager, F.V.M.P.A.
100. Oldham, Robert W., Manager, Lucerne Milk Company, Vancouver, B.C.
101. Omdahl, Sverre, Director of Agriculture, State of Washington.
102. Owen, Walter S., Q.C., Counsel, Ice Cream Manufacturers' and Distributors' Society of British Columbia.
103. Palitti, Walter A., Managing Director, Shannon Dairies Limited.
104. Palmer, William, President, Merchants' Independent Co-operative Association, Victoria.
105. Pappas, John, Manager, National Dairies Limited.
106. Park, William J., Dairy-farmer, Pitt Meadows.
107. Peterson, August, Dairy-farmer, Agassiz.
108. Page, Wyvern M., Dairy-farmer, Matsqui Prairie.
109. Reifel, Harry F., Dairy-farmer, Milner, and General Manager, Bellavista Farms Limited.
110. Richardson, William T., Dairy-farmer, Chilliwack.
111. Rive, Charles, Manager, Comox Co-operative Creamery Association.
112. Robin, Samuel J., Manager, Arctic Ice Cream Division, F.V.M.P.A.
113. Robinson, Henry G., Secretary-Manager, Vancouver Island Dairymen's Association.
114. Ross, Miss Rosamond H., Nutrition Consultant, Metropolitan Health Committee, Vancouver.
115. Rundle, Arthur D., Dairy-farmer, Camp River District.
116. Savage, Dr. Kenneth G., Dairy Inspector, B.C. Department of Agriculture.
117. Seldon, George F., Assistant Manager, Purdy's Café, Vancouver.
118. Shaneman, Jack A., Dairy-farmer, Cowichan.
119. Sherwood, Ernest G., Dairy-farmer, Lulu Island.
120. Schroeder, John R., Consumer, Chilliwack.
121. Spencer, Dr. Leland, Professor of Marketing, Cornell University.
122. Sprott, Alderman Mrs. A., Representing City of Vancouver.
123. Smith, Dr. Frederick W. B., District Veterinary for B.C. Section, Federal Control.
124. Smith, Robert J., Dairy-farmer, Ladner.
125. Stevenson, Archibald, Dairy-farmer, Cowichan.
126. Tomer, Mrs. Mabel L., Chairman, Agriculture Committee, Vancouver Local Council of Women.
127. Tuson, William G., Manager, Honeydew Restaurants, Vancouver.
128. Vonesch, Joseph, Dairy-farmer, Upper Sumas.
129. Wakelyn, Donald W., Milk Inspector, City of Victoria.
130. Walls, Charles E. S., Secretary-Manager, B.C. Federation of Agriculture, Victoria.
131. Watts, William B., Managing Director, Watts Marketing Research, Vancouver.
132. Wardrop, William, Former Dairy-farmer, Deroche.
133. Webb, Arnold E., Inspector, B.C. Milk Board, Victoria.
134. Weller, Ernest F. G., Managing Director, Hazelwood Creamery Limited.
135. Wells, Gordon E., Dairy-farmer, Sardis District.
136. Whittaker, Mrs. Margaret, Dairy-farmer, Duncan.
137. Williams, Trevor J., C.A., Vancouver.
138. Wilson, James W., Executive Director, Lower Mainland Regional Planning Board.
139. Wood, William, President, Richmond Milk Producers' Co-operative Association.
140. Withington, Arthur E., Dairy-farmer, Nicomen Island.
141. Wyndlow, George, Producer-Vendor, Yellow Point, Vancouver Island.
142. Zanatta, Moses, Dairy-farmer, Abbotsford.
143. Zink, Leonard A., Dairy-farmer, Chilliwack.

List of witnesses, Clyne Commission, 1954-55.

reversed the decision made by BC courts and declared the issue to be outside the boundaries of provincial jurisdiction.

Efforts to regain control of milk marketing were compounded by the effects of the depression hitting the Fraser Valley market and by the refusal of the Independents to appoint a member to the Milk Board in late 1930. In 1932, Berry's *Dairy Sales Adjustment Act,* too, was declared *ultra vires:* invalid on the grounds that it lay outside the boundaries of provincial jurisdiction. When a meeting was called to "dispose of almost $17,000 collected in levies, no producers attended to lay claim to their share — a powerful indication of the resentment felt by the cooperative members."[11]

These initial legal skirmishes laid the groundwork for much that would come over the next 20 years. The *Natural Products Marketing Act* (1934) was an attempt to develop a way that federal and provincial marketing boards could work together, sharing responsibility for a jurisdiction. Fruit growers and milk producers were the first to draw up schemes for approval.[12]

Relative calm in the industry marked the era of the Second World War, a time of prosperity supported by a strong demand for products and by producer and consumer subsidies. In 1942, authority for the marketing of milk was transferred to federal control under the Wartime Prices and Trade Board and prices to both producer and consumer were fixed. The federal government relinquished control in 1946 and the Provincial Milk Board, created under the *Public Utilities Act,* took control of both producer and consumer pricing. By 1948, with wartime demand curtailed, increasing surpluses

meant the renewal of the old marketing war. According to Clyne, it was "carried on with a vigour which was not only unreasonable, but also unprofitable to the producer and unbeneficial to the consumer."[13]

During these years, the Milk Board's right to set pricing was at constant issue in a series of legal challenges. By 1953, the Milk Board was not only being openly defied in terms of pricing, but was attempting unsuccessfully to manage the growing surplus. It introduced a quota system based on distributor pools. Because independent distributors sold much more milk on the fluid market than did the FVMPA, the average fluid milk quota of independent producers was much higher than that of cooperative producers. This quota system made plain the income disparities existing between the two groups of producers.

The industry Clyne stepped into as he began his commission was an industry in deep difficulty, regulated by a governing body that the majority of producers — on both sides of the debate — had little trust in. "At first blush," Clyne wrote, "it might be said that the farmers of the Fraser Valley and their Vancouver distributers, both independent and cooperative, are the authors of their own misfortunes, and that they might very well be left to stew in their own juice. The answer, however, is not as simple as that. It is very much in the public interest that there should be a continuous and adequate supply of wholesome fresh milk to the metropolitan area, and such a supply is imperiled if the farmer is exposed to disastrous losses occasioned by destructive competition even though that state of affairs may be occasioned by the activities of the farmer himself and

Ex 367

DEL-EDEN farm
R.R.2 LADNER, B.C.
PHONE 423-M
REGISTERED GUERNSEYS

MURRAY DAVIE

Mar. #1/55

Your Lordship:-

I wish to submit for your consideration several personal observations, that I believe have a bearing on the milk industry in this area:-

(1) It is doubtful whether the primary producer is receiving his full and just share of the consumers dollar, spent on the purchase of fluid milk.

(2) It is doubtful whether every producer of this primary product in this valley area has an automatically lawful right to his proportionate share of that market, unconditionally. Rather, I feel that the quality of his product, the availability, and consistency of his supply and delivery, together with the efficiency of his marketing outlet, are a more natural and logical assumption. To a certain degree, the business he enjoys is that which he earns though the various manipulations of his more desirable product, in the hands of those entrusted with the processing and ultimate delivery of his product. This same principle applies, largely in the field of general business, or the numerous professions on which we are so dependent for economic progress and livelihood.

DEL-EDEN farm
R.R.2 LADNER, B.C.
PHONE 423-M
REGISTERED GUERNSEYS

MURRAY DAVIE

(2)

(3) It is doubtful whether the milk producers in this particular area can be expected to form any united action as a producer group, on a voluntary basis. Whether this fact is unfortunate, or otherwise is debatable. That which may appear logical in theory, is often proved unsound in actual application. Monopolistic tendencies are not looked upon with favor by either, the general public or, government bodies. Those of us who desire the benefits of fair competition when we purchase, should be expected to extend the same privileges to others, whom we depend upon to create a market for our own products. This does not mean necessarily, that controls are not needed. Economic stability of an industry, or nation, depends to a great extent on sound democratic authority at the different levels of business and government.

Your Lordship, as I see it, the Royal Commission which you head, is concerned with over-production of milk as far as fluid consumption is concerned, and the other aspect of the possibility of diminishing returns to the primary producer. I admit that I am not qualified to advise on the processing and distributing fields of the milk industry. However, as a producer, I

DEL-EDEN farm
R.R.2 LADNER, B.C.
PHONE 423-M
REGISTERED GUERNSEYS

(3)

take this opportunity to submit a few items of thought for your consideration:-

(1) Perhaps the policy and method of inspection of farm premises could be applied more efficiently. I do not imply that a clean premises automatically guarantees a high quality product. It is reasonable to expect that many farmers are enjoying a share of the fluid market, that would otherwise be denied, under a more practical method of supervision.

(2) Perhaps, the growing tendency of farmers, to use milk substitutes for calf rearing, is not in the best interests of the milk industry at the present time. In some instances, the policy may be practical, but in the main, widespread use would be a questionable practise.

(3) Perhaps, the immediate establishment of an organization known as the Milk Foundation would be of material benefit to the industry, and the national interest. The activities of such an organization would be to educate and advertise milk, and milk products, as essential food elements. A strong tangent would be to investigate the possibility of supplying more milk to schools, particularly

---

DEL-EDEN farm
R.R.2 LADNER, B.C.
PHONE 423-M
REGISTERED GUERNSEYS

MURRAY DAVIE

(4)

in the low income districts of the population. Also it would arrange for proper displays of material at larger exhibitions, and other merchandising activities. Primary producers may wish to have representation on such and share in the financial responsibility. While a limited national policy is now in effect, I can see little benefit to our own local immediate problem.

(4) Perhaps it would be sound policy to have government officials, conduct the test for butterfat content of all milk, at the various delivery points.

Your Lordship, I thank you for your consideration in allowing me time before you. Like all producers, I look forward to your Royal Commission Report, confident that it will be most progressive.

Signed:-

Murray Davie.

Submission by Murray Davie of Del Eden Farm to the Clyne Commission, 1955.
BCARS MICROFORM

Marginal Operators walk out

I think that probably the best thing that ever happened to the dairy industry was the Clyne Royal Commission. A lot of farmers, quite a few of them marginal operators, walked out at that time because they were so upset with some of the regulations that were brought in. It was a good thing they did because they weren't contributing much anyway. They were no good for themselves or for the industry.

Allan Toop, 2000.

his distributors."[14]

Clyne's recommendations fell into four broad categories: legislation, qualification, equalization, and pricing. Legislation pertaining to the production, processing, and distribution of milk and milk products was scattered over seven provincial jurisdictions. He proposed consolidating all legislation under a single statue, the *Milk Industry Act,* for reasons of "convenience, simplicity, and certainty, and also for the purpose of securing co-ordination in administration."[15]

The second category of recommendations had to do with health and safety and the qualification of producers to participate in producing fresh, safe milk of high quality for public consumption. Clyne proposed the inspection and grading of farms for quality of milk, animal health, and for maintenance of facilities, and gave the Milk Board authority over penalties for non-compliance. Clyne also suggested that "only those farmers who conform to the public health standards set up for the production of milk for fluid consumption, and who produce milk of the same good quality in accordance with those standards, should be licensed to ship to the fluid market."[16]

The third category of recommendations had to do with equalization between qualified producers who would share the higher priced fluid market under standards set by the government and through a quota system. Clyne proposed the creation of a Milk Marketing Board, essentially a clearinghouse for milk, whose duty was to see that "the unqualified producer is not permitted to ship for fluid consumption, and that the fluid market is divided fairly among all qualified produc-

ers."[17] When its authority came into effect on March 1, 1957, the Milk Board "took possession of all milk sold to fluid distributors and processors, pooled the proceeds, and divided them between all producers in proportion to their shipments, irrespective of whether the milk of the individual producers was sold for fluid or manufacturing purposes."[18] To keep the equation in balance, Clyne recommended that all distributors be required to pay the same price for the same products, namely the producer price fixed by the Milk Board.

The fourth category of Clyne's recommendations had to do with pricing. Consumer price controls, according to Clyne, had "already proved to be a failure."[19] He believed in price controls to the producer and suggested that formula pricing should be the means of setting the producer price. Formula pricing "should recognize changes in producer costs, the purchasing power of money, and a supply-demand adjustment factor."[20] ■

D. Brown
Cross-Exam.

back to where the grass grows.

(Witness aside)

LEONARD ARTHUR ZINK, sworn:

EXAMINED BY MR. NORRIS:

Q   Mr. Zink, you are a dairy farmer?

A   Correct.

Q   Where is your farm?

A   About six miles west on the Trans-Canada Highway, of Chilliwack.

Q   How long have you been working?

A   Well, I was born in that farm.

Q   On the same farm?

A   Correct.

Q   It was operated previously by your father?

A   Right.

Q   Do you operate that yourself?

A   Yes, correct.

Q   Your father shipped his milk where, when he operated it?

A   Several years ago he shipped to the Guernsey Breeders in Vancouver.

Q   Where do you ship now?

A   I operate two farms -- three farms, to be correct, and we have two outlets; we ship to the Chilliwack Dairy in Chilliwack, and the Palm Dairy in Vancouver.

Q   Your total shipment, first of all, with the Chilliwack Dairy, speaking generally, on average?

A   Well, we average 18 to 20 cans.

Q   And to Palm Dairy?

E.C. Carr,
Cross-Exam.
(by Mr. Wills)

their share of the total of that sale.

MR. WISMER: I see. I think that is all.

CROSS-EXAMINATION BY MR. WILLS:

Q   Mr. Carr, I understand that the Milk Board obtains its authority to operate from the Public Utilities Act. Is that correct?

A   That is correct.

Q   We were told by Mr. Norris earlier that under a recent decision in the Avalon case, that the public utilities act was to protect the consumers only?

A   Yes, that was the judgment.

THE COMMISSIONER: That was so held.

MR. WILLS: Q In practice, do you follow that procedure, in your operation of the Board? In your operation of the Milk Board, you tried to serve three masters, as it were, the producer and the consumer and the distributor, is that correct?

A   That is correct.

Q   What particular service does the Milk Board render to the primary producer?

A   We establish the price which he will receive for the milk supplied to a milk dealer.

Q   Is it true that under licences granted by you the primary producer can only deliver milk to one named distributor?

A   No.

Q   All right. I should have qualified my question.

Page of transcript of Clyne Commission in which Leonard Zink, a farmer in Chilliwack, is examined, 1955. BCARS MICROFORM

Page of transcript of Clyne Commission in which Mr. E.C. Carr, chairman of the Milk Board, is cross-examined, 1955. BCARS MICROFORM

CHAPTER VI

Marketing Milk —
Living up to Change

SODICA promotes its milk. Here, Milton Parker of Royal Dairy is sending a carton of NOCA milk by plane to the Penticton V's, a hockey team taking part in the World Hockey Championships in Dussledorf, Germany. According to SODICA press releases, this was the first time in history that fresh milk was transported such a great distance — over 5,000 miles. The milk arrived in good shape. The team won the championship, 1955.

LLOYD DUGGAN COLLECTION

J.V. Clyne's Royal Commission on Milk came to the dairy industry after at least 35 years of acrimonious competition in the Fraser Valley between independent producers and distributors and the FVMPA. Clyne's recommendations were specific to the Fraser Valley. The application of Bill 70 did not immediately apply to other areas of the province — the Peace, the Vanderhoof area, the Bulkley Valley, the Okanagan, and the Kootenays — although some areas (such as the Bulkley Valley and the Peace) eventually created their own quasi-legislative system with the power to regulate regional quota based on Milk Board policy. The consolidation of legislation, coupled with the Milk Board's ability to enforce that legislation, meant that many small or marginal producers — who could not or would not live up to the new stringent regulations monitored by the Dairy Branch of the British Columbia Ministry of Agriculture — were edged out. While the last of the small shippers disappeared in the 1960s and early 1970s, overall milk production in the province continued to rise. This was also an era in which processors and producers from all facets of the industry worked together to promote the sale of milk and milk products first under the banner of the Milk Foundation, and later under the BC Dairy Foundation.

The post-war era also marked the beginnings of a real shift in consumer buying habits. The emerging centrality of the supermarket gradually began to ease out home milk delivery. When in 1940 a supermarket in Ohio introduced the gallon jug of milk, its appearance on supermarket shelves and its enthusiastic reception by consumers was judged to be important enough to be included as a feature in *Butter-Fat* magazine, although the commentary suggested that such a thing would never fly in British Columbia

Other changes were also afoot in the industry. Safeway, under the brand name Lucerne, began processing milk and semi-fluid products for sale in Safeway stores. It relied on 18 big shippers, beginning its operation with bulk tank pick up only. Other dairies became increasingly reluctant to pick up small quantities of milk on stands at the roadside as they had done since before the turn of the century. Refrigerated milk tanks became mandatory in 1972, allowing farmers to rapidly cool their milk, while enabling dairies to streamline milk pick up. Canada's switch from the imperial system of measurement to the metric system in 1976 required further modifications both on the farm and at plant level.

The British Columbia Milk Board, created in 1956 under the *Milk Industry Act,* has been altered over the years. In 1990, the British Columbia Milk Marketing Board (BCMMB) was constituted — as other commodities had been before — under the *Natural Products Marketing Act.* It was created "for the purpose of promoting, controlling and regulating the production, transportation, packing, storing and marketing of milk, fluid milk and manufactured milk products within British Columbia under provincial authority and for the purpose of regulating the production for marketing, or the marketing, in interprovincial trade, of milk, fluid milk, and manufactured milk products, under federal authority."[1]

Today, the Milk Marketing Board's existence is

British Columbia Federation of Agriculture

536 BROUGHTON ST.
VICTORIA, B.C.
TELEPHONE EV 3-6020

CHAS. E. S. WALLS
MANAGER

P.O. BOX 907

M I N U T E S   O F   D A I R Y   M E E T I N G

HELD IN ROOM 218, SKYLINE HOTEL, RICHMOND,

DECEMBER 14th, 1960.

THOSE PRESENT:

Armstrong Cheese Co-operative Assn.   John Fowler

Fraser Valley Milk Producers' Assn.   J.J. Brown, H.S. Berry, T.J. Robertson,
                                             J.C. Brannick, Gordon Park.

Island Farms Dairies Co-op. Assn.   D.D. Chapman, W.W. Michell, J. Looy

Mainland Dairymen's Assn.   W.C. Blair, R.J. Barnes, W. Savage,
                                             F.V. Bradley

Richmond Dairies   Les Gilmore, Jack Savage

Shuswap-Okanagan Dairy Industries
            Co-op. Assn.   Everard Clarke, E. Stickland

Vancouver Island Dairymen's Assn.   A. Stevenson, Ted Thompson, W. Taylor
                                             R.J. Purdy

B.C. Federation of Agriculture.   Chas. E.S. Walls

The meeting was opened at 10.00 a.m. Mr. A. Stevenson was unanimously elected as
chairman.

1. PROPOSED AMENDMENT TO SECTION 57 OF THE MILK INDUSTRY ACT.

   The chairman first called on the BCFA Manager to read a two page document
   covering the proposed amendments to Section 57 of the Milk Industry Act
   (Appendix "A" attached).

   The meeting then decided to review each paragraph of the suggested wording for
   Milk Board Orders, as covered in paragraphs (a), (b), (c) and (d) on page 1
   of Appendix "A".

   (i)   It was moved by W. Savage and seconded by T.J. Robertson that paragraph (a) now be amended to read:

        "The original cost of the raw product which shall be the price as
        established by the Milk Board for each area as the producer price
        for each classification of milk based on formula, plus any other
        authorised charges which shall constitute the vendor cost of the
        raw product; plus".
                                                          Carried.

   (ii)   It was moved by W. Savage and seconded by John Looy that paragraph (b)
   be amended to read as follows:

        "Processing cost, which shall be the average production cost of each
        dairy product within each controlled area, as established by figures
        supplied to the Milk Board by all processors of dairy products within
        each area. Such cost shall include (a) processing and manufacturing;
        (b) container and packaging supplies; (c) selling and advertising;
        (d) administration and general; plus".
                                                          Carried.

BC Federation of Agriculture minutes, 1960.

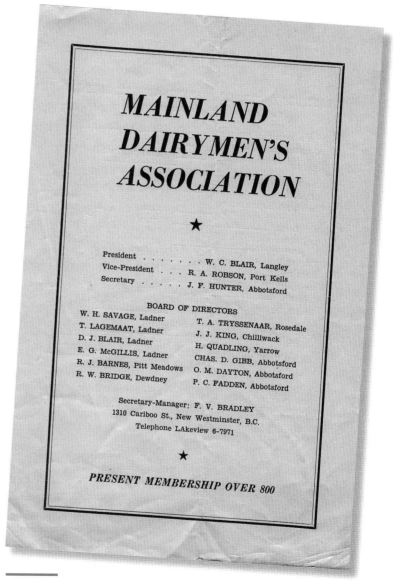

MAINLAND DAIRYMEN'S ASSOCIATION

★

President . . . . W. C. BLAIR, Langley
Vice-President . . . . R. A. ROBSON, Port Kells
Secretary . . . . . J. F. HUNTER, Abbotsford

BOARD OF DIRECTORS

W. H. SAVAGE, Ladner   T. A. TRYSSENAAR, Rosedale
T. LAGEMAAT, Ladner   J. J. KING, Chilliwack
D. J. BLAIR, Ladner   H. QUADLING, Yarrow
E. G. McGILLIS, Ladner   CHAS. D. GIBB, Abbotsford
R. J. BARNES, Pitt Meadows   O. M. DAYTON, Abbotsford
R. W. BRIDGE, Dewdney   P. C. FADDEN, Abbotsford

Secretary-Manager: F. V. BRADLEY
1310 Cariboo St., New Westminster, B.C.
Telephone LAkeview 6-7971

★

PRESENT MEMBERSHIP OVER 800

Mainland Dairymen's Association pamphlet, 1959.

Increasing Harmony

assured and its power is consolidated under a package of related provincial and federal legislation. The *Natural Products Marketing Act* empowers the BCMMB to regulate the production of all milk — both fluid and manufactured — within the province of British Columbia. It can allot quota to producers, issue licences, and collect levies required to underwrite its work. While it functions as a body independent of government, its work is related to government through a "superboard" overseeing all commodity marketing boards and performing both an appeals function and a governance function. The Milk Industry Advisory Committee (MIAC), a group composed of representatives from all facets of production and processing, fulfills an advisory function to the Milk Marketing Board by reviewing subjects related to pricing, production, regulation, and supply, and by advancing recommendations.

Because the *Natural Products Marketing Act* limits the jurisdiction of the BCMMB's powers to within the province of British Columbia, federal legislation is required to govern and to protect the export of BC milk to other provinces. The *Agricultural Products Marketing Act* does this job by regulating the interprovincial movement of fluid milk — milk for consumer use in liquid form (including cream). Like the *Natural Products Marketing Act,* the *Agricultural Products Marketing Act* empowers the BCMMB to allot quota, to issue licences, and to collect levies to support the work of the Board.

Industrial milk is milk surplus to the needs of the fluid market — including milk being processed into semi-fluid products such as ice cream, yogurt, and cottage cheese, as well as milk being processed into butter,

skim milk powder, evaporated milk, and all cheeses. It is regulated under the *Canadian Dairy Commission Act.* Passed in 1966, this act set up a crown corporation, called the Canadian Dairy Commission (CDC), to delegate authority to the Milk Marketing Board to act on its behalf to issue licenses and collect levies "to control the production of industrial milk and to generate the funds required to cover the cost of exporting surplus products."[2]

The creation of the Canadian Dairy Commission was designed to provide for the production of sufficient industrial milk to meet the domestic butterfat requirement of Canadians and to tender for sale on the world market the resulting surplus of skim milk powder. Initially, the CDC allocated production quotas to each province based on historical industrial milk production data. BC received an allotment, or Market Share Quota (MSQ), of approximately three percent of the Canadian total, with penalties imposed for overproduction.

As sales of milk and semi-fluid products increased in BC, less and less true industrial milk was available for the production of butter, skim milk powder, evaporated milk, or cheese in manufacturing plants throughout the province. The CDC eventually recognized this dilemma facing the BC industry in general and the FVMPA in particular, and introduced a production ratio of 65 percent fluid and 35 percent industrial milk.

Marketing Milk

In the early 1960s, most of players in the dairy industry agreed to begin working towards the common goal of increasing milk sales through consumer education. The

The Vancouver Milk Foundation's mini barn, a travelling exhibit of farm animals used as part of the Milk Foundation's mandate to educate people about nutrition and dairying, late 1960s.
SANDRA MAY COLLECTION

Sandra Holdsworth, later Sandra May, at one of her first appearances as representative for the Vancouver Milk Foundation, 1961.
SANDRA MAY COLLECTION

Speaking With One Voice

We were competing with each other and with other processors for a share of the same market, the same customers. We would win some and lose some. Somehow, Neil Gray managed to convince these competitive individuals that there were times in a highly regulated industry such as ours that it would be desirable that processors speak with one voice and sometimes sit together as one group and listen to voices from other parts of this country. As a result of his efforts, the BC Dairy Council was formed in 1973.

Frank Bradley, 2000.

first Milk Foundation in the province was the Victoria Milk Foundation, organized in 1959 by Frank Norton of Northwestern Creamery. Two years later, the Kootenay Milk Foundation and the Vancouver Milk Foundation were formed, with H.S. Berry as the Vancouver branch's first president. However, because the Vancouver foundation was the only one with an office and a full-time home economist, it became the provincial base for fluid milk promotion. A not-for-profit society, its mandate was to educate in the fields of health and nutrition — especially stressing the "health-giving value of milk and milk products."[3] It served the Lower Mainland from Hope to Horseshoe Bay dispensing nutrition education without brand advertising. Its board contained three producers and the managers of local dairies: Jersey Farms, Palm, and Dairyland.

In 1961, the Vancouver Milk Foundation hired its first paid employee, Sandra Holdsworth (later Sandra May), to develop free public programming to encourage the sales of fluid milk. May was creative and developed a variety of promotions: demonstrations at the Pacific National Exhibition and at conferences and trade fairs, farm tours with children, and plant tours with nurses and home economists. Activities included the distribution of educational materials for teachers, nurses, dieticians, dentists, and doctors; providing assistance with school-based nutrition projects; and consumer education through radio and television. Farm tours with classes of grade three children proved to be extremely popular. According to a column by Edith Adams in the *The Vancouver Sun* on June 30, 1965, farm owners welcomed these classes. They "let the children walk calves, play with newborn kittens,

see milking machines, and even let some ride pigs, a sight that brings tears of laughter to onlookers. It's rewarding to receive letters from grateful principals, like the one who said, 'in one class, only two of twelve children had been to a farm. In another class, 24 of 37 children had never visited a farm.'"

Until 1969, only producers shipping milk to Palm, Jersey Farms, and Dairyland were contributing to the Vancouver Milk Foundation, but in that year, the provincial government developed a promotional program whereby all producers in the province were required to contribute. Levies were imposed based on each producer's volume of fluid milk sales. A Dairy Products Promotion Fund Committee was appointed by the provincial government and consisted of five producers representing different regions of the province, one from Vancouver Island, two from the Fraser Valley, one from the Okanagan, and one from northern BC. Its first producer members were G.W. Park, S. Hanson, J. Pendray, W. Martins and W.L. Blair. Processor representatives were J.H. Argue, Foremost; H. Weins, NOCA; and W.J. Aird, Dairyland.

In 1973, the processors of BC formed the BC Dairy Council with Ev Crowley of Avalon Dairy as its first president. Membership included Canada Safeway, Dutch Dairy Farms, Foremost Foods, Fraser Maid Dairy Products, the Fraser Valley Milk Producers Association, Island Farms, Palm Dairies (Nelson, Prince George, Vancouver, Victoria), Shuswap Okanagan Dairy Industries Cooperative Association, Silverwood Industries (Vancouver and Victoria), and Little Mountain Dairy. Its primary role was to provide a liaison

As part of its community outreach, Dairyland sponsors the television show, *Reach for the Top*. Here, Neil Gray is seen congratulating a contestant. Champion skiier and guest presenter, Nancy Greene stands in background along with host, Terry Garner.

between the processing segment, the Ministry of Agriculture, and the BC Milk Board. One of the first acts of the new council was to guide the industry through the metric conversion process.

The new BC Dairy Council members agreed to join with producers to form the Milk Foundation of British Columbia. The directorship of the Milk Foundation was based upon share of funding, and included the five producer members of the Dairy Products Promotion Fund Committee and three processors from the Dairy Council. In its first year of operation, the Foundation provided 214,000 brochures, booklets, and posters to 996 schools in British Columbia.

Shortly after the formation of the BC Milk Foundation, Jack Gray was hired to manage its affairs, and under his leadership the role of the Foundation gradually expanded, eventually incorporating the coordination of the activities of the Dairy Bureau of Canada within BC. In the early 1970s, the name of the Milk Foundation was changed to the BC Dairy Foundation to reflect its role in the promotion of milk in all its many forms. The BC Dairy Foundation works through advertising, nutrition education programs, and promotional campaigns to increase the consumption of milk and milk products in the province. Today, the Board is comprised of six producers (two from the Mainland Dairymen's Association and four from the BC Milk Producers Association), three processors, and two ex-officio members (one from the BC Milk Marketing Board, and one from the BC Ministry of Agriculture, Food, and Fisheries).

From the 1950s, the Dairy Farmers of Canada (DFC) collected levies from producers across the country. These funds were used for lobbying the government. The Dairy Farmers of Canada subsequently formed a promotional group, the Dairy Bureau of Canada to develop and manage marketing programs using funds generated by levies from its members. In 1970, the Bureau initiated a "dairy case training program" for the retail sector that was very successful, not only in the training of store employees, but also in directing attention to the contribution dairy products were making to the overall profitability of the store. In 1994, the DFC merged with the Dairy Bureau of Canada to form one organization under the name Dairy Farmers of Canada. Today, the DFC marketing department promotes products like ice cream, cheese, and butter. Its nutrition department ensures that health officials and consumers receive useful information about nutrition.

The National Dairy Council of Canada was formed in 1919 to provide a forum for the dairy processing industry throughout Canada. The Council, located in Ottawa, had notable success in creating direct links with all federal agencies and continues to be instrumental in the development and maintenance of a variety of programs for the betterment of all segments of the dairy processing industry. The industry in BC has supported the Council for many years: two members from BC, Neil Gray and David Coe, have served as chairmen of this national body.

Introducing Regulation in the 1950s

Dr. A. Kidd, 2000.

At the time of the Royal Commission I was assistant livestock commissioner, assistant chief veterinary

George Okulitch, general manager of Dairyland, and Sandra May of the BC Milk Foundation, Vancouver, late 1960s.

Sandra May appears on KVOS television to promote milk as part of healthy nutrition, late 1960s.

Dairyland promotes its products through its sponsorship of Storybook Farm at the Pacific National Exhibition, 1969. Margery Gray is working at the booth.

Carr Retires

With the retirement of Milk Board chairman Mr. E.C. Carr, the industry has lost a friend of exceptional ability and devotion to its welfare.

John Hulbert, "Report of the Dairy Industry to the 36th Annual Convention of the BC Federation of Agriculture," 1969.

inspector, and commissioner of fur farms. There were three groups that started the farming in the Fraser Valley. The original people were the English, Irish, and Scots people. They were the ones who broke up the land and started farms. By the time I came back from overseas, the Mennonite people had taken over at Abbotsford and Matsqui. The English, Scottish, and Irish weren't too successful in keeping their sons at home. Part of the reason was that they were living on say, 500 chickens and 11 or 12 milk cows. The dairy people would freshen their cows in the spring and they would dry them up in the winter and in the Fraser Valley they'd run out of milk in the winter season. These people, a lot of them on these small farms, were living in borderline poverty. This is one of the reasons why I've been in favour of marketing boards, so that there can be control that guarantees a good product to the consumer and also guarantees a living to the farmer.

I should tell you that back in the 1920s, the farmers of the province were complaining that there were no veterinarians. At that time there were very few veterinarian in BC — maybe a dozen in the Fraser Valley, and one or two on Vancouver Island, and maybe a couple in the Okanagan. That was about it, and so the politicians of that day changed the Milk Act to require veterinarians only to inspect the barns and milk houses. That proved to be no help at all because there weren't enough veterinarians around, so they would just go around the barns and count windows. So this was the situation and we got up to a stage in the Fraser Valley where there were over 10,000 people that were shipping milk or cream at some time during the year.

The umbrella over all of this was the FVMPA. They were the ones that took in all of the excess milk. Opposed to them was the Mainland Dairymen's Association and the Independent Dairymen's Association. These people had most of the good shippers, and they were just shipping to the fluid market, and that left FVMPA holding the bag. I noticed in the Royal Commission on Milk *that the FVMPA sold at one time 50 percent of their milk on the fluid market; whereas the Independents were selling over 90 percent of their fluid milk on the market. So the big, big problem was what to do.*

Borderline dairy farmers existed and there was no regular good supply of milk during the winter. This was the way it was up until during the war. On September 1, 1942, milk marketing came under control of the Wartime Prices and Trade Act. *This act kept peace in the Valley during the war because it controlled the producer price and subsidy. So things went well during the war, and then the federal government got out of this June 30, 1946. The provincial government amended the* Public Utilities Act *to establish a Milk Board on July 2, 1946, which established producer prices. On October 1, 1953, they abandoned producer prices and the whole situation in between 1946 and 1953 was very poor. There was overproduction and there was no help for farmers. It was a very difficult time for everybody in the milk business, so on January 31, 1954, under the* Public Utilities Act, *the government established a quota for each person shipping on a 184 day basis. The quota at that time was to be of no money value and could not be sold.*

When the honourable J.V. Clyne wrote his report, he

MILK is the Chief Dependable Source of Lime which is Essential for Building Sound Teeth

The kind of teeth a baby develops depends upon his mother's diet during pregnancy and the nursing period.

A Quart of Milk Daily is required by the mother and later by the growing child to supply sufficient tooth building material.

Teeth a Living Part of the Body

Like all other body parts teeth depend upon food for growth and strength. Milk is richer than any other food in tooth-building materials.

Mother's Responsibility

A child's teeth begin to develop in the third month of pregnancy. At birth these first teeth are fully formed in the jaw. Previous to birth the mother is the sole source of food for her baby. Her diet must protect her own teeth and build those of her child. The daily quart of milk, liberal amounts of fruits and vegetables, together with cod liver oil or daily exposure to sunlight, supply excellent tooth-building materials.

The Permanent Teeth

The permanent teeth are already forming in the jaws of the baby at birth. The first permanent tooth erupts at about the sixth year, the last one appears near the end of the growth period. It is evident then that tooth-building material is needed throughout the entire period from birth to mature life.

The mother's diet during the nursing period and after that the child's own diet, can best be fortified with a quart of milk daily.

The quality of an adult's teeth depends largely upon his diet.

A Pint of Milk Daily is recommended as the minimum amount to maintain healthy teeth throughout adult life.

The Adult's Teeth

Contrary to former theories, there is now convincing evidence that an adult's teeth may be made and kept strong and sound by proper diet even though they have previously shown definite signs of deterioration.

Proper Diet Will Build Strong Teeth and Keep Them Sound Throughout Life

Early brochure from the National Dairy Council of the United States, 1920s.

Bringing About Understanding

The Dairy Committee meetings continued to benefit from the presence of the Dairy Commissioner to discuss changes to the BC Milk Act regulations and Dairy Branch regulations. The Dairy Committee appreciates the continued discussions it had with the Milk Board and the staff at the Department of Agriculture. It is through such frank, thorough discussions of our problems that a degree of understanding is attained whereby such a committee can bring about improvements for the well being of the producer.

S.W. Cross, "Report of the Dairy Committee to the 39th Annual Convention of the BC Federation of Agriculture." 1972.

suggested, first, that we had to enforce inspection of dairy farms. Only farms of a certain quality could be shipping milk to the market. Secondly, we had to eradicate brucellosis (this is what they called undulant fever in humans). Undulant fever in people was quite common at that time and there was no test even of raw milk. Raw milk dairies had no idea of how good their milk was and there were many raw milk dairies at that time. I believe when I moved to Vancouver Island in 1955, there was something like 32 raw milk dairies alone in the Victoria area.

I took over in Victoria on January 1, 1956, after the Commission was over. We were authorized to set up a committee to make up the regulations as to what we should require. It was a big board to make regulations, but we were the core that sat down and put it all together: Dunc Mackenzie was the assistant dean of agriculture at UBC under Dean Eagles; John Kirkness, from Chilliwack, was a dairy farmer; and myself. All the regulations came under the new Milk Industry Act.

After the regulations were made, I was authorized to hire laymen to do the dairy farm inspecting and so we took 22 of these to UBC and we lectured them for a month and took them to dairy farms to teach them how to properly enforce regulations and also how to get along with people. Before I moved to Victoria, I went to Danny Nicholson who was president of FVMPA at the time, and asked him how to get this job done. It was a big one. He said, "Well, Abe, if you can treat everybody the same, you'll have no trouble with us. Treat everybody the same." Of course that was the key.

We trained these 22 men, and 12 of those, I remem-

ber, went into the Fraser Valley to start as dairy farm inspectors. The other 10 inspectors were stationed at other locations in the province. I was talking to Mr. Kiernan, the Minister of Agriculture, and he said, "Abe, we'll back you always, as long as you give fair warning." So we put the fair warning right on the inspection sheet, and I must say it was amazing. We did have a minimal amount of problems and when a few of them did reach Mr. Kiernan, he stood behind us all the way. It worked out very well for us. When it was all over two years later and we had inspected everything in the province, there were just over 1,700 dairy farms — out of the thousands of people who had been shipping — that were qualified to ship milk. The rest of them just quit shipping milk.

Tuberculosis was pretty well eliminated by this time on dairy farms. So the next thing we had to do was to establish a program to get rid of brucellosis. The federal government supplied the vaccine and we administered it in my branch, the Veterinary Branch. Brucellosis was a bacteria and it was very costly to farmers because it caused abortion of calves. Mr. Kiernan asked me one day, "What are we going to do about this vaccination, Abe?" I said, "Well, I've been in Alberta, and I've been in Ontario, and in those two provinces the provincial government pays the veterinarian a dollar a calf to vaccinate. I would like you to approach the BC Veterinary Association and offer them a dollar a calf." He says, "Well, how much would that be a year?" I said, "Well, it would be in the neighbourhood of $70,000 — I've done my homework on that." The idea is to build up a "wall of immunity" before you start testing

168

The Effects of the Clyne Commission

Neil Gray, 2000

The first few years after 1956 were difficult economic times for most British Columbians, including most dairymen, thus change was not immediate. As the licensing of farms proceeded, many marginal producers unable to meet the new regulations imposed by the Dairy Branch left the industry. Refrigerated bulk tanks started to appear in the Fraser Valley in 1957 and 1958 and installations accelerated over the next five years. The switch to bulk at farm resulted in the FVMPA purchasing tanker transport, and it wasn't long before those producers unable to go bulk were forced to leave the industry.

Prior to 1956, we had at least 3,000 active shipping members, but by the time can-hauling was discontinued, we were down to less than half that number. Notwithstanding this reduction in numbers, the volume of milk being produced by those still operating increased quite dramatically. Today, less than 800 producers throughout the province produce an ever-increasing volume, fully meeting the needs of the fluid market, as well as providing our provincial share of industrial milk, that is, milk being processed into butter, ice cream, yogurt, cottage cheese, mozzarella, and cheddar cheese.

Formula pricing, as recommended by the Clyne Commission, was eventually introduced by the Milk Board, and when in place, the price paid to all licensed producers was protected from inflationary pressure. The formula utilized published cost indices for all farm input costs, and did indeed keep pace with the actual cost of production at farm. However, the processor buying price, as also established by the formula, increased in a corresponding fashion. Without price control at the retail level, and as a result of intense competition, the processor margins became badly eroded. Inflation was rampant in the 1970s and to help offset low or non-existent margins, increases in the volume of throughput became necessary for survival. Consequently, many smaller, marginal dairies were unable to compete, eventually disappearing from the marketplace. Such a trend continues to this day.

The changes as introduced by the Clyne Commission have created an element of professionalism at farm level second to none in Canada, and today we have in BC a guaranteed supply of top quality milk meeting the needs of large, well equipped and dedicated processing facilities which are, in turn, providing the consumer with milk products of exceptional quality at reasonable prices.

Income Assurance for Agriculture

Income assurance was an enormous change to the dairy industry in this province, and it was brought on by Dave Stupich of the NDP. He stood up and spoke for agriculture and took the position that farmers have every right to make a reasonable living.
Jim Waardenburg, 2000.

the older animals because once they test positive, they go for slaughter. So, we set up a meeting with the council of the BC Veterinary Association. After a little bit of small talk, Mr. Kiernan said, "'Now, I want to talk to you about this calf vaccination. We need to get all the calves between five and eight months of age so the vaccination titre would be gone by the time they go for blood tests. I'd like to make a deal with you people to get your association to vaccinate all the calves for a dollar a calf." The president stood up and said, "Oh, that's fine. We can do that." That's how quick it happened. The veterinary association did a tremendous job of getting the vaccinations all done and this program allowed veterinarians to settle in Prince George, Smithers, the Peace River, and the Kootenays because it guaranteed them help on their basic income. It worked very well. The brucellosis eradication included compulsory calf vaccination, testing to find infected animals, and establishing brucellosis-free areas. The Health of Animals Branch of Canada's Department of Agriculture eventually took over the whole brucellosis program in Canada and we were completely free by 1975.*

Support for Farmers
Barrie Peterson, 2000.

We moved to British Columbia in 1943 when my dad bought this farm and started in the dairy as a member of the FVMPA. Two or three years after that, he became president of the Agassiz local. He was still president in 1956 when he was killed in a tractor accident. At that time, I took over the farm. I shipped milk until the end of March, 1999, when I turned it over to my son, Gordon. We've been on the farm here now for 57 years. Dad presented a brief to the Clyne Commission on behalf of the Agassiz members of the FVMPA. At that time, they numbered about 90; now, we're down to about 40. I had been president as well as secretary of the Agassiz local and, in 1971, I became a director of the FVMPA.

We have seen a great many changes. When we started in 1943, we had no electricity here on the farm. Milking was done by hand. It wasn't until 1947 that we got electricity and put in a milking machine. We saw the change from milk cans to can coolers, then finally to bulk tanks. We've seen production go up. Throughout the last 40 years or so, we've seen a very rapid change in the facilities for milking cows, and we've also seen an increase in the quality. The quality bonus program instilled a pride in most of the farmers in BC to put out the best quality they could. The standards at one time were a million somatic cell count. Now they're way down. The very dramatic decrease in somatic cell count meant an ultimate increase in milk quality. There were tests all along for quality, but there's no way that the standard 25 years ago was as high as it is today. The axiom around our farm is "if it isn't good enough for me to drink, it's not good enough to go in the tank." Quality sells products. You've got to hit top quality. I think that's what we try to do at the farm level and at the processing level.

In 1972, I was put on the BC Federation of Agriculture Dairy Committee. I was chairman of that when the NDP government put in the Interest Reimbursement Program and the Income Assurance

Minister of Agriculture, Ken Kiernan (left) turning on water at O'Keefe Ranch near Vernon as part of ministry-sponsored irrigation program to improve grasslands. Mr. O'Keefe is watching, 1953.

LLOYD DUGGAN COLLECTION

Minister of Agriculture, Ken Kiernan (left) and Dean of Agriculture at UBC, Blyth Eagles, share a piece of NOCA cheese, 1950s.

LLOYD DUGGAN COLLECTION

Program for the dairy industry. We were the first to get a program finalized and a lot of the other commodities followed. There were a number of us heavily involved in the decision making at that time, working with government to make, I guess, a lasting impression on the dairy industry. There were a lot of people involved: we had representatives from all areas of the province, Vancouver Island, the North, the Peace River, and the Okanagan. The program put a lot of dollars back into the dairy industry and turned the dairy industry throughout the province right around. There's one thing that has to be recognized whether you like the NDP government or not: when Dave Stupich was Agricultural Minister, he brought in income assurance and interest reimbursment — some of the biggest benefits that agriculture has received. I've criticized Stupich for the things he's done and whatever he hasn't done, but he had the ear of the Premier, Dave Barrett. Ken Kiernan and W.A.C. Bennett brought in the Clyne Commission, and Stupich brought in income assurance. I think they were two of the political factors that we have all benefited from. Income assurance heralded an improvement in the income of dairy farms at that time. It allowed people to improve their facilities and to improve their production and quality.

My father joined the FVMPA when we came from the prairies. We were members of the Saskatchewan Wheat Pool, which was a cooperative. I guess he felt strongly about cooperatives. The other factor was that the people that we bought the farm from were shipping to the FVMPA and the majority of farmers in this area at that time were shipping to them.

When we first moved here, we milked 16 cows by hand. As we got into machines, we slowly increased. Today we milk about 50 cows, give or take a few. We have never gone larger. I guess I was away too much of the time. We never had full-time hired help. It was a family operation. We depended on family to handle it. Our transition from Guernseys to Holsteins was very gradual. When Gordon finished university, he got some purebred Holsteins. He just gradually worked them in. A year ago, we transferred the quota to him. That's the way things changed. We started here with 40 acres of cleared land and that's more than doubled right now. What has also happened is that production per acre has gone up substantially at the same time. Not only production per cow, but production per acre. I think this is true for the whole Fraser Valley. We improved the land. I'm like a lot of people. I want to leave it in better shape than when I took it over.

People Make the Difference

Jim Waardenburg, 2000.

Back in 1968, my dad and brother were farming on Nicomen Island on a rented farm. They had the rented place for seven years, and prior to that were in Hatzic for one year. Before that — from 1955 to 1960 — they farmed in Mount Lehman, next to Danny Nicholson's farm at the end of Mount Lehman Road. I still have some of the income tax returns from those days when Dad farmed there and milked about 25 cows. I think the gross income for that year was less than six thousand dollars and he had to pay for everything out of that and live as well. Things have certainly

A PROGRESS REPORT OF THE

Vancouver Milk Foundation

2940 MAIN STREET, VANCOUVER 10, B. C.

On June 14, 1968, your Board of Directors held the seventh annual meeting of the Vancouver Milk Foundation.

At the meeting the following slate of officers were elected for the 1968-1969 fiscal year:

President	Neil Gray
Vice-President	L. Gilmore
Secretary-Treasurer	H.D. Burbidge

Directors: Messrs. G. McKay   J.C. Cherry
L. Gilmore   R. E. T. Irwin
G. Fawcett   R. E. Mitchell

The year end statement as audited by Thorne, Gunn, Helliwell & Christenson shows receipts from producers and distributors of $49,961.00 plus earned interest of $1,851.00 making total revenue for the year $51,812.00. Our expenditures for the year were $51,493.00, a total excess of revenue of expenditures of $319.00.

The Milk Foundation teaching aids are based on Canada's Food Guide; and since the Guide gives suggested amounts of milk to drink to help meet the average daily nutrient requirements for different ages, Milk Foundations are promoting their product while performing a worthwhile service. All nutritional material comes under the scrutiny of Dr. Beaton through the associated Milk Foundations. Due to his guidance the material is interesting, accurate and up to date.

Since the nature of the material is educational, our first objective in promoting good nutrition is by distributing the teaching aids to the school districts serviced by the Vancouver Milk Foundation. The area of service is determined by the producer members. The Vancouver Milk Foundation area covers seventeen school boards. These are Fraser Canyon, Agassiz, Mission, Chilliwack, Abbotsford, Maple Ridge, Langley, Surrey, Delta, Richmond, Coquitlam, Burnaby, New Westminster, West Vancouver, North Vancouver, Vancouver and Parochial. The three departments, Home Economics, Science and Guidance were separately contacted in the one hundred secondary schools, and all 496 elementary schools were contacted. Three hundred fifty-nine of these schools ordered 137,150 separate items. These items are workbooks, pamphlets, teaching guides and posters. Due to the demand for these teaching aids and the limitation of our budget, it has been necessary to place restrictions on quantities that may be ordered.

Brochure, Vancouver Milk Foundation, 1969.

TELEPHONE 657   P.O. BOX 533

VALENTIN DAIRY

DEALERS IN
HIGH GRADE MILK AND CREAM

PRINCE RUPERT, B. C.   Oct. 31, 1935

Mr. Wm. Billeter,

Smithers, B.C.

SKEENA CREAMERY BUTTER :: SUMMIT ICE CREAM

1.	408			
3.	408			
5.	510			
8.	408			
10.	408			
12.	510			
	2652	4%	.70	74.28
15.	408			
17.	408			
19.	510			
22.	408			
	1734	4.2%	.70	50.98
				125.24
88 qt. cert. sold=				
220# @ 4.1% @ .30				2.71
				127.95

All Accounts Due and Payable on the 15th day of each Month.

Invoice from Valentin Diary in Prince Rupert to William Billeter of Smithers, 1935. Valentin took milk from nine Smithers-area farmers to his plant in Prince Rupert despite the challenges of an erratic train schedule and of no refrigeration for cans en route. According to Billeter, "We learned that if we took really good care of our milk before it left the farm, it would take quite a bit of punishment. That meant lots of ice – and putting up ice was a chore in itself."

Bulkley Valley Dairymen's Association

The dairy industry began to be of importance in the Bulkley Valley during the 1920s with the establishment of a creamery and production of butter at Telkwa. The shipment of fluid milk to market in Prince Rupert developed during the 1940s. Production of fluid milk for the markets of Prince Rupert, Terrace, and Kitimat expanded during the 1950s and dairy farms located mainly in the Bulkley Valley became an important part of the local economy.

The production of milk was well suited to this area, but the problems of processing, marketing and distribution plagued the local dairy industry. In 1968, the Bulkley Valley Dairymen became a part of the Fraser Valley Milk Producers' Association with the FVMPA purchasing the processing facilities located in Kitimat.

Jim Davidson, "Brief Submitted to the Select Standing Committe on Food Study Costs at Smithers," August 23, 1977.

Les Gilmore hosts a school tour at his farm in Richmond, late 1960s. SANDRA MAY COLLECTION

changed. Dad wanted to retire and my brother didn't really want to farm by himself, so he approached me and asked me if I wanted to go in with him. I worked at the bank at the time, but decided that would be a good opportunity to start a different career, so we started looking for a farm to purchase. We were fortunate enough to find this place in Matsqui, and to get the dollars together to buy it. It's just over 200 acres. My younger brother decided to quit his job and come in on the farm as well, so we had a partnership of three brothers. We farmed there until the first of May 2000. Over the last number of years, I've been operating the farm with two nephews as full partners and that's

worked well. It's a good area to farm. We've done well here. As long as you've got good land, everything else will follow.

The herd size has changed enormously. We started out with about 35 cows and it's been growing steadily ever since. We left it with around 200. The number of animals increased, the quality increased, and production per cow has also gone up. When we came to this farm, it had a 1,000 gallon tank in the dairy. That was an enormously large tank for its day. But it wasn't very many years later that this milk tank was too small and we bought a 2,500 gallon tank. It lasted us about 20 years, then we replaced it with a

174

Interior Creameries plant in Telkwa, 1940. This creamery originally opened as the Telkwa Creamery in the 1920s and was bought by Interior Creamery of Prince George in 1939 to handle cream from the area between Burns Lake and Hazelton.

Bulkley Valley dairyman Jim Davidson after addressing FVMPA meeting, 1969. Association members, left to right: Jim McLellan of Glen Valley; Harold White of Dewdney; Jim Davidson; John Fennellow of Chilliwack; and Ed Smith of Courtenay.

Representing Farmers

At this time, I would like to express my sincere appreciation to the members of the dairy committee who gave so freely of their time and talents to accomplish the results we have this past year. The producers of this province should be as proud as I am of the knowledge and qualifications of the members of this committee that they appointed to represent them.
Barrie Peterson, "Report to the BC Federation of Agriculture Annual General Meeting," 1976.

5,000 gallon tank. So, it's gone up in size considerably.

Dad was a FVMPA member on Nicomen Island, so that automatically transferred onto this farm. When I became a full-time farmer, I applied for membership of the FVMPA and I've been involved ever since. It's important that you help direct the industry and participate where you can. I was always involved at the local meetings and producer meetings dealing with the dairy industry. It's been a good experience.

The Clyne Commission was a real help to all the dairy farmers in British Columbia. It made the market available to all the producers, not just a few. When income assurance came in, it was basically an undertaking by the provincial government to support agriculture to maintain a level of income for farmers to provide a stable return on investments and labour. Once we got into an area of getting reasonable levels of income, farmers started to upgrade their housing, facilities, bought more quota, expanded production and replaced equipment. Things in the middle 1970s were good for the dairy industry.

I think it wasn't until the early 1980s when the interest rates started to skyrocket that we got pulled down a peg or two, and those who had large loans — and probably a lot of the dairy farmers did because of the expansion that they had gone through — were having to pay the high interest rates. There were some desperate times in those days, but it was a lesson that people have to learn the hard way periodically. So, the industry learned — but so did the bankers — an expensive lesson in economics.

In the 1990s came a bit of belt tightening. The production controls were tightened, the revenue really didn't keep up with the increases in cost and again we had to expand in order to basically break even. Everybody is pulling their belts a little tighter; I don't think that's bad at all. It's just something that we have to get used to. In the 1990s, we've seen an escalation in farm consolidations. A lot of people have gone out of the business, and a lot of farms have doubled or tripled in size. There really aren't too many farm operations now that are the same size as they were 10 years ago: instead of milking 50 cows, they're now milking 150 cows, all forced by economics. Small farms will continue to survive as long as the original owner continues to operate it. There are still quite a few farms milking 50 or 60 cows and they're doing very well because they have no debt. But the minute they turn that farm over to the next generation or to anyone else, it will create a huge amount of debt, and that particular party will have to generate a whole lot more money from that farm operation than the original owner.

Whatever has happened at the farm level has happened in exactly the same way at the processing side. You have to have bigger equipment and more volume to compete. The FVMPA and the cooperatives from the prairies merged into one operation now known as Agrifoods International Co-op Limited. So, the Fraser Valley Milk Producers became part of a larger organization. Agrifoods International Co-op owns all of the assets and it owns all of the trade names like Pacific, Armstrong, Dairyland, and Alpha. The Agrifoods merger was dictated to us by the

Moving Milk, 1960s to 1990s

Keith Miller, 2000

I was raised on a small farm in Chilliwack. That was my background. We were one of the infamous two-can shippers. I got to know of the FVMPA at a very young age. It was our bread and butter when we were kids. My father went to work for the dairy in 1941, so I'm the second generation with the FVMPA. I started in the Sardis plant in 1951 and spent most of five years working in the cottage cheese area — I started washing gutters there. I worked my way through various positions until I became in charge of the tank truck operations in 1963. From 1990 to 1999, I worked as the regional transportation manager for Dairyland and then following mergers I became the corporate transportation manager, covering all of the Canadian provinces until my retirement in 1999.

At the Fraser Valley Annual Meeting of 1968, a resolution was passed to convert all shipping from cans to tank by January 1970. Because of inclement weather during the winter of 1969, that deadline was put back to the end of February 1970. On that day, all shipping to the FVMPA was by farm holding tank. At the end of December 1967, there were 1,307 farms shipping in the Lower Mainland area. As of March 1, 1970, 987 FVMPA shippers had converted to tank, the difference dropping out without conversion. In early March of 1970, Dairyland took over Jersey Farms' home delivery service, and with that 17 producers transferred to Dairyland for a total of 1,004 shippers.

Between 1958 and 1979, we went through a gradual upgrading of size of hauling trucks. The first truck ac-quired was a 2,200 gallon, and later we introduced 3,200 gallon trucks for greater efficiency. That was the largest truck we could go to on a tandem axle. In order to get greater capacity, the next step had to be to semi-trailer, and with semi-trailer, the next logical conversion was two sets of doubles so that we could pick up milk either as sets of doubles from the farm, or as individual semi-trailers as access dictated. One great advantage was that the milk was going directly from the farm to the processing plant. This provided a longer shelf life. All of this change was accomplished with great cooperation from the farms, and we got into semi-trailer farm pickup in early 1980. Actually, the first set of doubles was put on in August of 1979, but it operated on the interplant haul and gave drivers the opportunity to get familiar with the doubles before we actually got them working off the farms.

In the early 1970s, three major FVMPA plants remained in the Valley: the Sardis plant, the Abbotsford Delair plant, and the Burnaby Sperling plant. The Sardis plant was considered the plant of last resort and all of its production (other than cottage cheese and butter) went into powdered milk. The Delair plant was exclusively an evaporated milk plant. I think something that perhaps gets overlooked even now is that what we refer to as the dispatch office is actually the fluid milk control for the province. Abbotsford dispatch monitors and directs the movement of raw milk product throughout the province. This is no easy task.

marketplace. Competition in the marketplace dictates that the processing side be very efficient and that is what prompted the expansion into other areas of the country.

When we switched over to the delegate system, some of the farmers felt they had lost their voice. I would agree. I think a co-op is strongest when you have the full support of all the members and we had that in British Columbia. But once we started to merge with the other co-ops, there was just no way that 3,000 or more farmers could attend meetings and be involved with the numbers. In fact, the three co-ops, the two in Alberta and the one in Saskatchewan, were already on a delegate system and we had a partial delegate system already. Our annual meetings, for example, were open to all of our members, but a lot of members were in areas — like Smithers or Creston or Prince George or McBride — where members would have to travel a fair distance to attend, so they had representatives come to the meetings. I agree that the farmers have lost some of the direct contact. Farmers just don't feel as involved when they don't have a vote.

I was the last president of the FVMPA. It was a time that had come. There were mega-mergers taking place in the food industry and most everywhere else. A change of that magnitude is difficult to justify and to accept. We had seven on our Board, we had direct involvement on the part of our producers, and then we turned the control over to a new Board of seven people representing three former co-ops. I think our membership lost a lot. But I'm assuming that they would have lost more if we hadn't done it.

Now we have a say in a co-op that is two or three times larger. The way I think about it is that I was the last president of what used to be a successful co-op in British Columbia, but I was also vice president of the first amalgamated co-op that continued and has become stronger and is providing a stable home for milk. I think that's a tribute to the people that have run this industry for the last 100 years. It really is. We have a pretty solid home for our products, there is never a shortage of dairy products, and nothing ever gets dumped. There's a lot of critics out there. They should be thanking their lucky stars that we've got a solid, well run dairy industry in this country and in this province. I'm proud of the industry in total. We're in pretty good shape. There are going to be challenges, but if we can sit down as a co-op and as an industry and talk about it, we can find solutions and go on.

Certain people have been more involved in one area then they are in another. Barrie Peterson, for example, was one of the masters to get the Income Assurance Program in British Columbia. He worked right alongside Peter Friesen and a few others in this province and their efforts have had a huge impact on the dairy industry. Things don't happen automatically. You trace it back to people. I have a lot of respect for producers who have seen what needs to be done and then fought for it on behalf of producers. Everybody benefited. My area was more on the national side. The organization that I was involved in, the Dairy Farmers of Canada, has done a fabulous job of working to make this industry better. ∎

The Dairy Branch

Earl Jenstad, 2000

I got very much involved at the farm level. In 1972 all of the farms had to have bulk tanks. Once the bulk tank was put in, other changes had to be made to ensure that there was adequate turn-around room for the trucks. Significant changes took place when they started tank pick up with B-trains. They take a big turn around. Milk houses also had to change: they had to have reinforced floors so the tank was stable enough to be calibrated. The calibration had to be checked periodically. That created some interesting situations. I remember one farmer built a new barn and attached it to the milk house. The whole thing began to settle a bit, and it pulled the dairy down with it so that one end of the dairy had sunk. The farmer was losing big time. We've also had tanks delivered and adjusted, or the rods cut and re-welded. But overall in British Columbia, we have probably the best, most accurate system of measuring milk in Canada. Based on the work that we're doing here now, New Brunswick and Manitoba are looking at our system.

Mastitis used to be a kind of social disease, something that farmers wouldn't talk about. There was no really good method of measuring it. There was what they called the White Side Test, but it was very very subjective. When the somatic cell count came in, the science was established enough that somatic cell count could be related to the degree of mastitis. I was involved in establishing the mastitis program. Under this program, I had technicians going around the province taking samples from every cow, or every quarter in some cases, and sending them in for analysis. We got a tremendous amount of information and were able to establish a a somatic cell count program that everybody participated in. Initially we thought that 500,000 was a good cell count, and as time went on, farmers were making such tremendous improvements that

500,000 was way too high. Many herds are down to 30-, 40-, 50-thousand. That's from being able to track it and knowing what was happening on the farm that was causing it — whether it was malfunctioning of the milking equipment, sanitation of the equipment, freestall conditions, or bedding types.

At that time we felt milking equipment was one area that we really needed to work on. Herds were growing and milking equipment just hadn't kept up with the demands that were being placed on the systems. We developed standards for milking equipment.

We were constantly trying to find out new things and to keep in touch with research. My work was extension with a regulatory flavour behind it. Probably more than any other commodity in agriculture, we had a real pulse on who was out there because of the licensing system. We knew when farms came on stream; we knew when they left; and, because of the monitoring every month, in terms of production level and quality, we had a pretty good idea what was happening on the farm. I know some people didn't like that, but we tried to identify the problems and to help wherever we could. I know that's one of the parts of the job that I really enjoyed — to see real breakthroughs, to see problems being solved. In the last couple of years that I was with the Ministry, the extension people were told not to go on farms anymore because one-to-one contact with farmers was not considered efficient. Eventually you lose touch with what the farmer needs and wants. You just can't speak his language. Now there are only a couple of inspectors out there, and they don't have time to really stop and talk to the farmer. So I can see farmers would be missing that source of information. Some might be glad to be missing it, but by and large, the farmers I've talked to do miss it.

CHAPTER VII

Lives in Dairying

Beehive Ice Cream Parlour and Pool Hall, Campbell River, circa 1918.
BCARS C-04403

Houses were nearly all of logs, occasionally of split cedar, furniture was crude and often homemade, and nearly always limited to the barest necessities. Clothing consisted chiefly of working clothes; women's styles changed little, many made the same apparel serve for dress occasions for decades; and for seven or eight months of the year both men and women wore gum boots even to church or to dances. Roads were few and nearly impassible, except in summer, "so that the horses, instead of cantering along easily, had to wade and tug and pull ... When the streams were swollen by heavy rain ... it was necessary to draw one's legs on to the saddle, so as to prevent even long boots from getting filled. In going to church it was not uncommon for two persons to use one horse by the method known as "ride and tie," one riding a mile, tying the horse to a tree and walking on, while the other, having walked the first mile, rode the second, and so on.

John E. Gibbard, "Early History of the Fraser Valley 1808–1885," MA Thesis, UBC, 1937.

The stories people tell about their experiences developing their farms are many and varied. Some farms were carved from the bush or built up from bare land. Some were going concerns, handed down from a respected parent so that the stewardship of the herd and the farm would be in good hands. Along with plenty of hard work, early mornings, and late nights by the principal farm operator, the contribution of women and families remains an important feature of the family economy that makes the lifestyle and the business of dairy farming possible. Sometimes this work is easy to see — in the foreground in tasks such as milking, picking rocks, running machinery, working with animals. Sometimes it is in the background, especially in tasks like cooking, preserving, tending to a house and small children, keeping chickens or a large garden, keeping the books, or stepping in to milk or clean the barn to accommodate a husband's meeting schedule.

One common thread that links these stories is the knowledge that the need for adaptation and improvement is constant, both on the farm and in the policy and legislation that defines the bounds of the possible. Dairy farmers in all parts of BC have been proactive not only in pioneering better, cheaper, more efficient ways to do things on the farm, but also in living with — even thriving on — change. Farmers banded together before the turn of the twentieth century to form the BC Dairymen's Association to guide the growth of the industry, and now at the beginning of the twenty-first century, they continue to support a range of associations. For example, regional representatives of the BC Milk Producers Association, or the Mainland Dairymen's Association routinely participate in debates relating to present realities and future visions of the industry. The point is that change is constant: just as soon as one thing is fixed, more needs to be done. As Jim Davidson of Smithers says, dairy farming "is always about building up."

The Beginnings of Avalon
Jean Crowley, 1994.

It started with my father-in-law. The Crowleys came from Newfoundland in 1906, travelling by train. There were four children and one other child born on the way here. They arrived in 1906 on St. Patrick's Day (they were of Irish extraction) and bought or rented a house around the 5,000 block of Wales Road, north of 41st Avenue. That house has since been replaced. They had six cows and the use of a pasture down on Kingsway. That's how they survived while this present home was set up.

They had not been farmers previously. Grandfather Crowley had worked in a foundry. He didn't like it. He had done a little private farming in various places where he lived, but in St. John's there wasn't much other than their own garden. He seemed to enjoy life in Newfoundland, enjoyed skating, and was a good skater and belonged to the Loyal Order of Foresters and became president. But he felt confined there, so he made about four trips — a couple to the mainland and two to the West Coast — to see the possibilities.

Before he was married, he took one trip out West and worked in a coal mine and also in logging on

E.A. Wells with one of his prize cows, Exhibition Grounds, Victoria, circa 1910.

The Crowleys in Vancouver

Once settled, the Crowleys bought a horse and wagon and, with the help of a Sikh labourer, cleared the trees opposite the farm with powder and capstan, a device used with horse and pulley to uproot trees. They drained the land by digging a ditch, installing home-made drains, and broke up the ground with a team of horses. The whole process took about six months but the costs of this operation were more than offset by the first crop of potatoes from the new land.

Jean M. Crowley,
The Story of Avalon
Dairy Ltd, 1996.

Vancouver Island and found there was lots of work. But he was not a big man and the work was too heavy for him. He liked the West Coast, so he got enough money together to come here.

For the delivery of milk in the immediate area — probably half a dozen houses — they used a dog and small wagon. The milk was in cans and was ladled out into whatever container the customer provided. Grandfather collected all different types of bottles whenever he could. He took extra care to see that the milk was safe and clean. By 1909 they were pretty well set up, had a milk house, and used milk cans. In those days anybody who had a cow and six customers was in the dairy business.

We Kept Building Up
Charlie Porter, 1999.

My parents came to Chemainus. The sawmill was in Chemainus and they knew the manager or the owner of the sawmill. They spent one year down at the head of the bay by the mill and then they bought a quarter section. Of course, in those days — the 1880s — that was before the railroad and it was pretty primitive living. By the time my dad got married there was kind of an upset in the family. It was a big family, and he ended up, I guess, with this land here and the barn was being built at that time. He built temporary rooms in the end of the barn, and that's where they spent their 60 years of married life, and that's where I was raised.

My dad planted quite a big orchard and he figured that would be kind of a retirement plan, but it didn't

work out that way. He had to work out at Chemainus when they were constructing the new mill after the fire in 1932 or something like that. He never worked in the mill, but was section boss on the highway. He always kept a few cows, but it wasn't very profitable because there was hardly any land. When I left school, my idea was to build up the dairy industry. The only ones that seemed to be able to spend any money were the ones who had retail milk routes, and so that's when I started building up. That would have been 1938. We kept building up as we sold a bit more milk.

Early Days on Westham Island
Dorothy Davie, 1994.

The Savage family first arrived in this area in 1886. They landed on December 16 at Westham Island. It was a beautiful day. I remember my grandmother talking about how beautiful it was after the long trip from Ireland. She came over with five little children, one girl and four boys. Her husband had lost his potato crop in Ireland and decided to emigrate to Canada. He told her to come with him or wait until he looked around first. She came with him. His brother was already on Westham Island — the Jim Savage family.

My father was the youngest child, just a year and a half. They came by boat to New York and then took the train to San Francisco and then went by freighter to Steveston, then a gas boat to Westham Island. There was a fish cannery at Westham Island at the time. My grandfather bought property by the cannery and my grandfather died two years later with the flu. His name

Canadian General Electric advertisement, 1945.

The Aylard Family

I graduated from UBC in 1925 and Harold Steves was one year ahead of me. Charlie Rive was in and out. He finished in my year, I think. My father was a mining man. He liked farming but didn't like cows too much. I always wanted to be a farmer and when I left university, I went to Beecher Bay to a farm there and kept chickens and had three or four cows. In 1930, I went out to north Saanich and started farming. I had about 25 cows. Bert Doney and George Malcolm, farmers in central Saanich, started Registered Jersey Farms and they wanted milk, so we shipped milk to them.

Arthur Aylard, 1993.

was Rowland. The two oldest boys, ages seven and nine, died the following March with measles.

At that time they had a few cows and chickens. My grandmother carried on. She made butter and the fishermen came in and bought her eggs and butter and milk. When grandmother died, the youngest son, Willie, took over the farm.

Hugh Savage was my father. He didn't farm on Westham for long. When I was three we moved to the house on Fairview Road. And that was a grainery. There was Hal and Bill and myself and then they had Jack. At that time, my father had been looking after the Post Office and the boat from New Westminster would bring the mail down to Westham Island. He married my mother, who was a Kirkland, and they ran the Post Office and had three children on Westham Island. That's where I was born in 1909.

I was the first person to cross the Westham Island bridge, the bridge from the Island to the Ladner area. In March or April 1909, Dad hooked up the horse and buggy to go for a ride. I was in an orange box in the front of the buggy. We went up to the bridge and the planks weren't all nailed down but the sides were on and there were a number of people there, so Dad said that if they would hold the planks down, we would drive over. So the horse went over and we circled around Ladner and then returned.

When we moved into the grainery, we had a table and used orange boxes. It was pretty primitive. Finally, Dad raised the roof and put bedrooms upstairs. When we lived in the grainery, we didn't have purebred cattle, but we did have a Holstein bull which my mother

feared. I remember Fred May was killed by a bull. The Mays lived on Lulu Island across from my Uncle Johnny.

I went through school in Ladner and then went for five years to UBC. My teacher in Ladner suggested I take a year at UBC and get my first-class certificate to be a teacher. My father rented property in Richmond and Bill and I took cows over there and milked them by hand all summer and shipped the milk to Spencers store in Vancouver. Dad bought a herd of 21 head from Vancouver Island. Then the acreage was planted in grain and had an excellent crop and Dad said he could afford to send me to university. The admittance fee was $125 and Professor King got me enrolled. I finished first year in 1926, and then returned a second year, a third year, and a fourth, and eventually graduated in 1930. Then I returned for an extra year to take teacher training. I had my certificate but there wasn't a school to be had looking for a teacher. It was the deep Depression.

I finished schooling in 1931 and helped Dad all summer doing farm work. The Jersey Farms office was opening up in December and Mr. Fleming came out and offered me a job in the office. I started on December 16 1931. We had 179 quarts of milk to deliver. I stayed with Jersey Farms for 10 years.

Stanhope Farm
Gordon Rendle, 2000.

My dad came out from England and he lived in Calgary for a year. He got a job as soon as he landed in Calgary with people by the name of Black, who were very very good to him. He was only a kid of 18 at the

Edenbank wins first prize dairy herd at New Westminster Exhibition, 1907. Mr. Trapp, the judge, is on horseback.

BCARS D-07172

time and he worked for them for a year on a farm. Then in the winter, he had to haul a load of grain into Calgary with a team and he got caught in a blizzard. He got his feet all frozen, so that was the end of Calgary as far as my dad was concerned. He came out here. When he got off the train in Victoria, the first person he met was Bob Mercer. He asked him if he knew the old City Dairy in Victoria and if he knew anywhere he could get a job. Mercer sent Dad out to see David Blyth. They ran a dairy operation right where the Oakland School is today. He went and got a job there.

Well, anyway, Dad worked there and, I think, the Blyths were only there for a year before moving to the Hudson's Bay property on Cedar Hill Road. It came out to Cadboro Bay. It took in the block there. He worked for Blyth for seven years before he started on his own. I remember some of the stories he told because Blyth had a farm on Kangaroo Road out there in Sooke. They used to have to drive all the heifers and dry cows to Kangaroo Road. Well, that was a pretty big chore, wasn't it? So that's how my dad got started.

After Blyth moved off that farm, Jim Turner moved onto it and then my dad got the place down at Cadboro Bay, just north of The Uplands gates. I can remember Dad saying it was 1914 that he did that. Stanhope Dairy Farms was registered in 1916. I remember Dad saying the only time that he ever missed shipping milk or delivering milk in Victoria was in 1916 when they had the deep snow and he tipped the sled over going out the gate and lost the milk.

Dad moved away from the Cadboro Bay farm. He bought the place on Richmond Road, near Mount Tolmie, in 1937. That was the old Dean farm, and at one time it took in all of what was the Normal School and all the airfields down there. All that belonged to Dean.

Dad delivered milk to Victoria all the time. That's how he met my mother! He was delivering milk when he worked for Blyth and I don't know whether my mom was sweet on him, or he was sweet on my mother, but she was always sent out to the cart. Well, I guess if they sent my mother out, they always got more milk. Things haven't changed. Dad delivered the milk all in Victoria and for 20 years he supplied the Jubilee Hospital on Richmond Road. When we started at the hospital, delivery was strictly by cans. They had a little dairy down at the hospital. Then, I guess, they tried to get conscious of bacteria counts and the milk was fine when it landed at the hospital. But by the time they had dumped it into porcelain jugs and taken it up all through the hospital, they lost track of whether the milk came in yesterday's milk or last week's milk. There was no way of telling. There was no way of knowing. So, they wanted it pasteurized. We actually started pasteurizing our milk because we had a five-in-one pasteurizer.

A terrific cream trade at that time went to the hospital. At least 15 gallons of cream a day went just to the hospital. Eventually, they did away with the little dairy down in the hospital. They had an elderly fellow in there looking after it and he would divvy up all the milk on the floors and I think he retired and so we made a change. We started bottling the milk and we used seven different coloured caps for the hospital.

Gordon Rendle, 1985.

The Blair Family

Bill's (W.C. Blair) grandfather, Thomas Culbert, came to this farm in 1881. He bought it from the man who had homesteaded it. He came by sailing ship from Ontario, round Cape Horn. He came to New Westminster and then up to Fort Langley. They used to walk or

Langview Farm, home of the Blair family, 2000.

K.J. WATT

take a horse from here in Langley to Fort Langley to shop — it's five miles. Bill's grandfather raised his family here. He lost his wife when his two sons were very young. He sold the farm to Bill's father, George Blair, in 1902. Bill's father and mother were married in 1897 and they went to Ashcroft and worked on ranches and made some money and bought the farm from his father-in-law.

This farm was not very much cleared prior to George Blair's purchase of it. He had his first registered Holstein cattle in 1921 but he had registered Jersey cattle before that. At first they would have cream and make butter and take it to the market in New Westminster once a week, which was a day trip in those days. Originally, the farm was 160 acres, a quarter section. Now it is 140 acres. In 1910, George Blair sold the farm when the BC Electric came through the valley and there was a lot of speculation. He sold it to a developer and part of the farm was sub-divided off on the south side. But things didn't go as well as expected and so grandfather got the farm back. We lost some land at that time. So he bought 40 acres at the back to make up for what he had lost, so it brought it back to pretty much what he had originally.

I remember independent truckers coming into your farm yard and picking up the milk cans from the milk stand. We would cool the cans in water. We had a Blue Ribbon milking machine and then we switched to Surge. My father, whose last name was Livingstone, had owned Hillside Dairy in Cloverdale before the Crawfords owned it. He started bottling milk in the early 1930s. We used to ship to Seal Kap. Ron Miller worked as manager there. He would take our milk there in an old 1947 truck. It was just on Fraser Highway, half a mile down the road from here.

My father originally started in Richmond on No. 9 Road. He had a round barn. My uncle, Alexander Ewen, built that barn with the help of my dad in 1893 and my dad leased it and farmed there until 1923 when he bought the farm at Cloverdale. The Steves were at Steveston, the Grauers came later, and the Savages were around there. Mays were there when my dad was in Richmond, and McKays came a little later. Alexander Ewen had the farm on No. 9 Road; it is called Ewen Road today. It's where the Lafarge Cement plant is now. He had one of the first fish canneries in British Columbia at the end of No. 9 Road. My father had come to Canada from Scotland to run the cannery because he had his steam papers. This was in the fishing business in the 1870s. Sometimes they would go up the river on the ice when it froze over to take their produce to the market every Friday.

Doris Blair, 1994.

190

Every day there was a different coloured cap, so they knew if it was a yellow cap, that would be Monday's milk. It was pretty simple and it worked fine. We were the sole producers for that part. But we did work with old Herb Shepherd pretty closely. If we were short, he always helped us out. After my Dad started delivering to the hospital, I think it went pretty well for us. I think most shippers were running about four dollars a hundredweight for milk and we were getting nine to the hospital. The only thing is, hospitals never had enough money. Sometimes we'd go for three months without getting paid by the hospital. And then we'd pass it off to the feed companies. We always got money, but we always had to wait for it.

There have been so many changes. I remember when I was a kid going to school, I used to catch the Blue Line bus on Cadboro Bay Road and Cedar Hill Road. That's where it came out to and it was four cents for me to go to school. Four cents for the bus to ride to Victoria!

It's kind of strange, though, because if we go back to the time when they started the milk quotas, we were the largest milk producer on Vancouver Island at 1600 pounds of milk. It's hard to believe, isn't it? It will be 44 years in March since we took over the place where Stanhope Farm is now, in Saanich, just off the Pat Bay Highway. Last year, Stanhope-Wedgewood Holsteins had eight nominations for All Canadians which is extraordinary. And they were runner up for the second year in a row for Premier Exhibitor at the Royal Winter Fair in Toronto. They had Grand Holstein and Supreme Champion Cow, Rainyridge Tony Beauty, at Madison, Wis-

Auction Sale

OF

Pure Bred and High Grade Holstein Cattle, Imported Pure Bred and Grade Clyde Horses, Roadster Horses, Farm Implements, Machinery, Dairy Utensils & Miscellaneous Effects

Under instructions from CAPT. JAMES ERSKINE, I will Sell by Public Auction on the premises, Church Road, Sea Island, about 2 miles from Eburne Post Office, on

Wednesday, Nov. 1st, 1911

at 10 o'clock a.m., the whole of his Pure Bred and High Grade Holstein Cattle, Imported and Registered Pure Bred and Grade Clyde Horses, Roadster and general purpose Horses, Farm Implements, Machinery, Hay and Miscellaneous effects, which consist of in part as follows :

72 HEAD DAIRY CATTLE—Of these 30 head are Pure Bred Holsteins, with papers, and are from the DeKol strain, the sire of a number of these being the King of Eburne, one of the best bulls in B.C., nine of these are five-year-old cows, dry, and two or three will be due to calve by date of sale; 6 three-year-old Heifers, milking and in calf; 4 two-year-old Heifers, in calf; 8 Heifers, one year old and under; 1 Cow, ten years old, milking and in calf. (A number of the young stock are from "White Prince Mechthilde," the dam of this bull was Alice Botsford, owned by Mr. J. M. Steves, of Steveston, which cow has proved herself the best animal in her class in B. C.) 1 Bull, Pete Je DeKol Butter Boy, No. 41733, H.F.H.B., bred by Amery Stevens & Sons, of New York. (These are acknowledged to be the most successful breeders of Holsteins in America.) 1 Bull "White Prince Mechthilde." (The sire of this bull is "Sir Canary Mechthilde" and his dam "Alice Botsford." The above herd have been carefully selected and no money has been spared in their purchase.) 20 High Grade Holstein Cows, in full milk, some of which have been bred; 3 High Grade Cows, due by date of sale; 19 two-year-old High Grade Heifers, all supposed to be with calf.

27 HEAD OF HORSES.—Span of Imported Clyde Mares registered, 3750 lbs., eight years old, as follows: 1 Mare, "Peystone Baroness," sire Lathness Barron, dam Rose of Peystone; 1 Mare, "Rose Wattie," sire Red Wattie; 1 yearling Horse Colt by Royal Diamond Jubilee; 1 sucking Horse Colt by Royal Diamond Jubilee; 1 sucking Horse Colt by Shipmate, imported from Scotland. (These three Colts should make excellent sires.) 1 Mare, with foal at foot, 1600 lbs.; 2 Mares, with foal at foot, from Tamboline's Hackney stud; 1 Mare, with foal at foot from Roadster Horse; 3 Colts, rising three, from Diabmont; 1 Colt, rising three, from B. C. King; 4 Colts, rising two, from Diabmont; 1 Stud Colt, one year old; 1 Aged Mare, 1600 lbs.; 1 Filly, rising five years old, 1750 lbs.; 1 Driving Mare, six years old by Christmas; 1 Pony; 1 team Mares, six years old, 3100 lbs; 3 five-year-old Horses, 1400 lbs. each.

PRODUCE—125 Tons of First-Class Hay.

IMPLEMENTS AND MACHINERY—3 four-inch tire Farm Wagons; 1 Dump Wagon; 1 M.H. Binder; 1 Horse Hay Rake; 1 Tedder; 3 Mowing Machines, 1 T. & W., 1 M.H. and 1 McCormack; 1 Single Disc Seed Drill, 18 spout, M.H.; 2 Horse Cultivators; 2 sets Scratch Harrows; 1 Turnip Seeder; 1 Turnip Plow; 1 Breaking Plow; 1 Gang Plow (Cockshutt); 3 Oliver Plows; 1 Buggy; 1 Democrat; 1 Road Cart, nearly new; 2 sets Double Work Harness; 2 sets Single Harness; 1 set Double Driving Harness; 1 set Express Harness; 1 set Plow Harness; 5 Water Tanks; 1 Milk Cooler; 1 De Laval Cream Separator; Hay Carriage, fork, rope and blocks complete; 1 Fanning Mill; 1 Fairbanks Scale, 1500 lbs.; 2 Tarpaulins, 20 ft. by 10 ft.; 2 Wheelbarrows; 1 Meat Safe; some household effects and a variety of miscellaneous articles.

TERMS—For sums of $25.00 and under, spot cash; over that amount cash or approved lien notes at 3 months, with 8 per cent. interest.

Columbian Co., Print.

T. J. TRAPP, Auctioneer

Sale poster, Captain James Erskine, Sea Island, 1911.

TOM ERSKINE COLLECTION

FARM LANDS
IN BRITISH COLUMBIA
For FRUIT, DAIRY or CHICKENS

We have them from 5 to 500 acres. If you want one with little or no improvements, or one combining a high state of cultivation with a beautiful home, we can supply either at very reasonable prices, with a small cash payment, balance on easy terms.

Real Estate in Vancouver is increasing steadily in value. Excellent profits are to be made by judicious investments. We have choice properties in all sections.

We also deal in Timber, Mines, Stocks and Shares.

Write us for full particulars.

BEATON & HEMSWORTH
(Members Vancouver Stock Exchange)

Phone 7221   329 Pender St.
VANCOUVER, B.C.

Ad for farm land in British Columbia, 1911.

THE BRITISH COLUMBIA MAGAZINE, MARCH 1911

Growing Up Near Enderby

We had a tough time getting a dollar or two in those days but we always had lots to eat as everybody had a large garden, their own animals to butcher and lots of cream, butter, eggs, hams, bacon, and some farmers had their own wheat made into flour at the mill in Armstrong. The main source of a few dollars was the monthly cream cheque plus selling some logs, poles, cordwood or fence posts.

Stan Wejr,
"Memories of Trinity
Creek Area in the 1920s,"
in Okanagan History,
50th Report, *1986.*

consin, and that is a really, really tough thing to try and win. It's about the same as winning the Stanley Cup — that is the category it would go into.

I've been in dairy farming all my life. I just did it because I loved it. I love good cattle. In 1950, I took the BC cattle back to Toronto and helped with the show string. I think we only had one of our own cows in there at that time, but the rest were BC. And we were very fortunate: we had Reserve Champion at the Royal, and we had five firsts. The cow that won was Frasea Lady Perfection Wayne. I can't remember my own birthday or my wife's birthday, but I do remember the cows.

Okanagan Dairying

Dick Graham, 1994.

My grandmother finished high school in Vancouver in the mid-1880s and after she graduated she became a school teacher in the Fraser Valley, very close to Aldergrove, probably in the early 1890s. She taught in a school on the corner of the Fraser Valley Highway and what is now 264th Street. They bought property with the idea of farming, but then the gold rush came to the Kootenays and my grandfather essentially said that there was no sense in staying there, that the place for the future was the Interior — that's where the gold was. So shortly after they were married they moved to a little mining community by the name of Beaton, at the top end of the Arrow Lakes.

When my grandmother was pregnant with her first baby, who happened to be my father, she went to Kamloops to have her baby so my father was born in

Kamloops in 1897. Then when my grandmother was pregnant a second time, she went to Revelstoke to have her baby, so my uncle Charlie was born in Revelstoke. My grandfather, I think at the time, was operating the hotel in Beaton and it was quite a bustling place in those days.

My grandparents faced the fact that they would never get rich goldmining, but something exciting came along at that time and that was fruit farming in the Okanagan (this would be around 1903). My grandfather developed this property in Vernon for a combination of dairying and fruit farming — primarily apples.

My father grew up on the farm and when World War I came along he went into the service and served in England and Wales but not in France. He married a girl in Wales on June 19, 1919, and I was born on March 23, 1920, in Vernon.

My mother and father returned to Vernon in late 1919 and my grandfather had a ranchman's home on the property and my parents moved into that house. It had no electricity, was probably poorly insulated, with no running water and no indoor plumbing. Where the house stood is now part of the city of Vernon.

My grandfather had a lot of trouble with a moth in the fruit. This was a big problem in the Okanagan at that time. So more and more, my grandfather switched over to the dairy business and produced milk and actually started delivering raw milk in the town of Vernon. It was decided my grandfather would produce the milk and my father would look after the delivery of it as a separate business and he would buy the milk from my grandfather. That started in 1920.

Dairy farm near Vernon, 1967.

Coping with High Water at Colony Farm

One fall, in either the late 1950s or early 1960s, the Alouette Dam got so much water in it, they opened the flood gates to lower the water. It came down the Coquitlam River and it flooded out the land we had on both sides. I knew that we were going to get flooded out and I had about a hundred head of cattle in a dry barn over on that side of the river, so I went over and I chased them out of the barn. There was a hillside there, and I shut them out of the barn so that they could get up on the hillside above the water. And the next day, I got a rowboat, and I rowed across the fields above the fences. The water was high above the fences. I went along with the rowboat and got over to the barn. The barn must have had 10 feet of water in it. Two cows had gone back into the barn. One cow got up in the manger with her front feet and could keep her head above water. The other one was way down the barn. The barn had a dirt floor, but there were old planks to pile hay and straw on. And this cow was standing on the planks and they were floating her up. I was awful sorry I never had a camera that time to take pictures of what was going on!

Bill Howe, 2000.

At that time there were probably a number of farmers that would be producer-vendors, but my father started the first business that was strictly involved in the bottling and distributing of milk and he had rented premises in Vernon to carry on this business. Then we moved off the farm and my parents bought a home in Vernon with enough property to build a proper dairy plant. This plant was built in the late 1920s. The building still exists.

The business really did prosper even during the Depression. The only big problem my father had at this time was collections and I remember the amount of time he had to do book work and going out and collecting.

The Toop family in Chilliwack

A conversation between father and son, Glenn and Allan Toop, members of the third and fourth generations of the Toop family to farm in Chilliwack.

Glenn Toop: I had part of the old home farm and then I bought 15 acres from an uncle, and that was …

Allan Toop: 1945. April 1945. The reason I remembered it was April was because I was still in school and it was just before the end of the war. I can remember pushing out a can of milk in a wheelbarrow out to the road to the milk stand. My dad's uncle was still living in a little house on the farm. He hollered out to me, "The war is over!" I was so excited, I started to run with the milk can. Then I went in and told my mom and dad that the war was over and that there would be no school. Mom said, "Maybe it's just a false alarm, better get ready for school." So I did and the bus came because the bus driver didn't know that the war was over. He

took us to school, and then we were sent home. That was April 1, 1945, when we took over that farm from your uncle with four cows.

Glenn: I bought this farm in 1950 from a Mr. Chadsey.

Allan: This farm was probably the first farm that was settled in this end of this area, by James Chadsey. This is all Chadsey land in this area, but the farm that our family came to was up the road here about a mile and a half. Even the Zinks bought from Chadseys. And then when a lot of the Mennonites came in here in the late 1920s and early 1930s, most of the land they bought was Chadsey land too.

Glenn: The BC Electric went through, then they had a way to get into Vancouver beside the old steamboat. Before that, the only way to get into Vancouver was by horsepower or on the boat. When the train went through, we shipped our milk to Vancouver on the train. We put it on the train at Lickman Road. Of course, I was only seven years old, so I don't remember too much. I remember them building the railway track, that was built with horses and clip scrapers. When I was a kid at that home farm, there wasn't too much cleared land. We had milk cows. It was so much different. That little bit of land that you put into crop, you worked with a team of horses and a disc about three feet wide. And the haying. They never started haying until the twelfth of July. That was a magic day.

Allan: It could be a rainy day like this. Or they could have two weeks of hot weather, and they wouldn't cut any hay till the twelfth of July. I don't know where that came from, but when you consider it, the weather is usually quite reliable. The hay would be high and

The Toop barn during the flood of 1948, Chilliwack.

The flood of 1948, Chilliwack.

Toop family milking camp during the flood of 1948, Chilliwack.

Milking crew, flood of 1948, Chilliwack. Left to Right, back row: Glen Toop, Ellwood Toop, Ken Toop, Wheldon Toop, Dad Toop, Ron Toop, Allan Toop, Les Toop, Gerry Toop.

Not A Job

It's not just a job, it's a way
of life on the farm.
Audrey Peterson, 2000.

coarse. Now we have three crops in by then.

Glenn: *And we didn't sit in a tractor, air conditioned, with a radio and television and telephone. We sat on a mowing machine with a steel seat. I've seen mosquitos so bad that you could hardly see through them, and I'm not kidding you, just millions of them! I've seen them take the team of horses out to cut hay and the mosquitos were so bad, the horses would lay down. They'd have to take them back in the barn and blanket them. There was no good spray then, only a contact spray. Then, as soon as you quit spraying, well the mosquitos would lay right on the horses again. They would take them in the barn, blanket them, and leave them in there for a while. At night, when they turned the cows out, they would light a fire, throw green grass on it, it would make a big smoke, and all the cows would gather in the smoke. It would keep the mosquitos away. It wasn't so bad in the barn once you sprayed the barn and killed off everything that was in there. I've seen us having our breakfast or meal with a little smudge under the table. Just a little pot. Just a little smoke, not too much, but just enough that it would drive the mosquitos away.*

When I was a kid, we started early in the morning, we milked the cows by hand, and then fed the calves, or whatever there was to feed. Then we'd go in, have breakfast, and go to school in Atchelitz about two and a half miles. Then we'd come home and change our clothes and go to the barn and milk the cows again. I started milking cows when I was six years old, and I had to be there. So, that was the job and you didn't think anything of it. We would have to go out into the

Glenn Toop, 2000.   K.J WATT

bush and get the cows in in the summertime. We'd go get them out of the bush and the swamp, and get just about to the barn when the heel flies would hit them, and away they would go right back.

We gradually cleared the place. We took hundreds of cords of wood out, piled it along the road, and then the stumps would rot pretty quick, maple and birch and then they would pull them out with horses and chain, the small ones that they could, and then sometimes they would let the Chinese have 10 or 12 acres to grow potatoes, and if the land needed a lot of work, they'd get it for two years. They would pack all the stuff, the roots and everything else, and put in the potatoes and then it was a good field after that. The Chinese sure did a lot of work. My mother did a lot of work too. I don't know just how she did it. You know, we had to have hired help all the time, and she'd cook for them all.

Allan: *My grandfather, his dad, died when his oldest brother was seventeen and the youngest was five months. So, my grandmother had a real job on her hands. It was very very difficult. And everything was done by hand: all the water was pumped by hand, washing was done with a scrub board then, I guess. Was it, Pop?*

Mission and District Ayrshire Calf Club, 1940s.

Family Matters

I have seen the dairy industry develop over the last 45 years. It's been a busy life for people on the farm, but I think it's been rewarding at the same time. I know we've had to work long hours and we've had to work every day, but it's had its rewards. I've had a lot of rewards — the opportunity to meet people from all across this province, and all across Canada. It's something that you can't put a value on. I couldn't have done the things I've done if I didn't have a wife and family working with me. It hasn't been always easy for them. Audrey has travelled with me as much as she can, but she — and, of course, both of my girls and both of my boys — has had to milk cows and to do all the chores when I was away. I was fortunate when the kids were small that we had good neighbour lads who would pop in to help out. There are not too many operations that you can consider successful if you haven't got a wife and family behind you.
Barrie Peterson, 2000.

Glenn: Yes. It was mostly overalls once in a while. When they got so stiff you couldn't get them on, you'd wash them. There was a ditch — running water — right alongside the house, and it had a box drain, and it had sides and then cleats on the bottom and cleats on the top, and running water. We would take our overalls and just hang them in there, tie them to one of these cleats and leave them overnight. They'd be nice and clean in the morning.

Did I tell you about the sturgeon that broke our barn up? That inspector came around after the flood of 1948, he didn't know a sturgeon from a coho. I had a small farm, and he had missed my place. So I went up to my brother's one morning and he was there. So, we kept talking, and he says, "Where do you farm?" I showed him, and he says, "Oh, I missed that place." He was checking the drinking water and everything, and I said, "The drinking water is good. I've been drinking it." Then he wanted to know if there was much damage done by the flood. I said, "Well, my barn was pretty badly smashed up." "Oh, what happened?" I said, "Well, a great big sturgeon got in there and somehow he got his head stuck in a stanchion. He just beat that barn to pieces." He would have believed me if it hadn't been for my brother starting to laugh. During the flood, there was water all around us, but there was just enough water to get in the barn into the gutters.

Allan: See, we weren't on this farm the year of the flood. We moved here later. The old timers' houses were all put on the high ground because they knew where it was. In the flood, the Toop families — there were several then — had about 100 cows between the different

families and we ran them all together and moved them to high ground the day the dyke broke. We stayed over there for two weeks and milked all the cows by hand, and there was quite a few of us to do it. But to me, it was an adventure because I was 17, but for my dad and his brothers, it wasn't such an adventure. When we came back it was messy, though.

Glenn: Stinky and messy and all the sludge. Then it turned blistering hot. High water would come up every year and flood all the low spots. There was one thing about it, you could catch nice trout in any little running stream there was. I remember one morning I didn't want to go to school. Pink eye was going around. So I dropped a hay seed in my eye, and of course it turned real red. As soon as my mother spotted that, she said, "You can't go to school today ... you have pink eye!" I went fishing, I'll never forget that morning. I got 11 nice trout. I bet you there wasn't a quarter of an inch difference in the length of them. There were fish in every little ditch.

Milking Cows Would Be The Thing To Do
Walter Smith, 2000.

I was born on the twenty-seventh of September, 1922, in Holland in a family of 10. When I grew up, the times became very poor in the Depression years of the 1930s. Jobs were very scarce and the living standard was poor. Two of my older brothers went to high school but the jobs which were available brought in no money to speak of. My dad would say, "They can't even earn their wash water," yet they lived at home. Being number five in the family, I figured out that milking cows

198

Souvenir Programme

★

Boys' and Girls' Club Work

Exhibits, Demonstrations, Judging, Etc.

at

CHILLIWACK, B. C.
September 13th to 15th,
1943.

"Upon the association and intimate alignment of the policy of the United States and the British Commonwealth and Empire depends more than upon any other factor the immediate future of the world."—Winston Churchill, June 30, 1943.

★

Vancouver Exhibition Association
John Dunsmuir, President

Chilliwack Agricultural Association
C. L. Worthington, President

Program, Boys' and Girls' Club exhibition, Chilliwack, 1943.
FRED BRYANT COLLECTION

Boys' and Girls' Department Headquarters, Marquee
CHILLIWACK EXHIBITION GROUNDS

Professor H. M. King, Animal Husbandry Department, University of British Columbia, Chairman.

Supervisor of entries—Miss Echo Lidster.

MONDAY, SEPTEMBER 13th

4:30 p.m. Supper.
5:00 p.m. Halter Making Competitions, Mr. T. G. Stewart in charge, Dominion Department of Agriculture.
7:00 p.m. Swine Carcass Judging Competition. Messrs. R. C. Trimble, Hector Ford and Harold Steeves, in charge, Dominion Department of Agriculture.
(This competition will be held in the store of David Spencer, Limited, through the courtesy of Colonel Victor Spencer.)

TUESDAY, SEPTEMBER 14th

8:00 a.m. Breakfast.
8:30-11:30 a.m. Grooming and Preparing for Competitions.
10:00 a.m. Swine Judging, junior exhibits.
11:30 a.m. Lunch.
1:00 p.m. Judging of the calves in the calf clubs.
Official judges in the open classes to officiate for each breed.
4:30 p.m. Supper.
5:00 p.m. Junior Stock Judging Competition.
Messrs. S. S. Phillips and G. L. Landon in charge, B.C. Department of Agriculture.
Two classes of dairy cows—Dr. J. C. Berry, University of B.C.; Mr. Oliver Evans, Chilliwack, B.C.; Mr. R. L. Davis, Vancouver, B.C.
Class of hogs—Mr. H. L. Ford, Dominion Department of Agriculture.
Class of horses—Dr. M. Sparrow, B. C. Department of Agriculture.
Class of sheep—Dr. S. N. Wood, University of British Columbia.

F. D. GROSS TROPHY FOR STALL COMPETITION
To be judged daily and points awarded; cumulative score to determine the winner. To be judged Tuesday, 9:00 a.m., to Wednesday, 1:30 p.m.

WEDNESDAY, SEPTEMBER 15th

8:00 a.m. Breakfast.
9:00 a.m. District Elimination Competitions in dairy cattle, swine, potatoes and poultry for Fraser Valley and Vancouver Island teams.
Future Farmers of America.
4-H Teams from the State of Washington.
(a) District Elimination Competitions under supervision of Mr. S. S. Phillips, Secretary, Boys' and Girls' Club Work, Victoria, assisted by Mr. T. P. Devlin, Winnipeg, Man.

   (1) Dairy Cattle (4 breeds)—
      Dr. J. C. Berry, University of B. C.
      Mr. Frank Clarke, Colony Farm.
      Mr. R. L. Davis, Vancouver, B. C.
      Mr. K. A. Hay, Vancouver, B. C.
   (2) Swine Competition—Mr. H. L. Ford.
   (3) Potato Competition—
      Mr. H. S. MacLeod, Dominion Department of Agriculture.
      Dr. G. G. Moe, University of British Columbia.
      Mr. T. A. Leach, New Westminster, B. C.
      Mr. Wm. Coell, Dominion Department of Agriculture.
   (4) Poultry Competition—
      Mr. T. Sommerville, Vancouver, B. C.
(b) Future Farmers of America and 4-H Clubs Competition—
      Dr. S. N. Wood, University of B. C., in charge.
Contestants to use same classes of live stock as the District Elimination contestants. Additional class of horses to be provided.

12:15 p.m. Lunch.
12:00-12:30 CBR Farm Radio Broadcast.
Mr. Fergus Mutrie, farm commentator, direct from the Exhibition Grounds.
1:30 p.m. Showmanship Competition—Prof. H. M. King, Animal Husbandry Department, University of British Columbia, in charge.
Senior
Intermediate
Junior.
Judges: Mr. Oliver Evans, Jersey Fieldman; Dr. J. C. Berry; Mr. W. H. Hicks, Experimental Farm, Agassiz, B. C.
3:15 p.m. Presentation of B. C. Stock Breeders Challenge Trophy to the winning team by The Honourable K. C. MacDonald, Minister of Agriculture, Province of British Columbia. To be followed by presentation of other prizes.
3:45 p.m. Stock Parade.
4:15 p.m. Supper.

Chief Richard Malloway

Chief Richard (Ritchie) Malloway of Yakweakwioose, was born in 1907 and recognized in 1971 by the Chilliwack and District Chamber of Commerce for his contribution to the community. He was interviewed by Butter-Fat in that year about his work on Dr. Knight's dairy farm and about the subsequent development of his own herd. He joined the FVMPA in 1926.

When I started at Dr. Knight's, I was so young, I wasn't able to milk cows, so my job was to feed and herd them. We used to herd the cows on the road in those days. During the time I worked for Dr. Knight, I admired the cows so much, I thought I would some-day have cows of my own. After Dr. Knight, I went to work for Charlie Evans. He was milking 70 Holsteins at that time. Between these two jobs, I managed to save enough money to buy three cows. I still had to

Chief Richard Malloway

work out, though, to help establish the farm. I spent several years working as a logger, but it kept me too far from home. I took a job with H.M. Eddie's Nursery which meant I was able to stay at home. I would milk the cows in the morning and then walk to work.

One of the greatest drawbacks or difficulties of farming on the reserve is that I do not have the deed to my land and I cannot get one. Because of this, I cannot borrow from the bank. I have no deed for my land; therefore I can't sell it. When it is time to retire, I have three possibilities to follow. The first is to sell it to a member of my band. The second possibility is to turn it over to one of my sons. They aren't interested. As loggers and carpenters, they work eight hours a day, five days a week and probably make more money than I do. I can't really blame them because dairying is a 16-hour a day, seven days a week job. I think the non-Indian boys are leaving the farms for the same reason. The final thing I can do, and I guess that's what I will do when the work is too much, is sell my cows and quota and what equipment I can and leave the land. But, I still have my health and my mind is good, so I think I'll stay at it for a few more years.

Butter-Fat, September, 1971.

would be the thing to do. It would provide excellent food and if I could learn to milk cows by hand, I could make some money. And that's what I did. When I finished grade 7, I was able to get a job on a farm. I left home for my future life.

To spend my whole life milking cows for somebody else was not my idea. While still in school I was told about opportunities in California. Then World War II started and everything was kept on hold for five long years. To immigrate right after the war was impossible. But then an application was granted to go to Canada. On May 7, 1948, I was on the boat Kota Inten, and landed in Quebec and by train ended up in Edmonton. Standing on the station with my suitcases I was met by the immigration field man who introduced me to the farmer who had sponsored me. My wages would be $50 a month plus board and room.

As a young man of 25 years with enthusiasm and high hopes, alone in a new country, new language, new customs, it was hard. What a different way of life! But soon I picked up some words, and that made it easier. My farmer also spoke German, an asset for communicating.

After four years of hard work on the farm, I developed bleeding stomach ulcers. That meant a change of occupation by necessity. I was able to get a job in the city of Edmonton. On my days off I would visit dairy farms in the area. In the meantime I met and married my wife in 1957. After some time we talked about going dairying to Vancouver Island and, as I had holidays in August of 1958, we spent some time in the Fraser Valley and Vancouver Island. We fell in love

Jersey herd in the Fraser Valley, undated.

Frank Norton Sr.

My father was born in 1884 and educated on Saltspring Island. He left school at an early age and worked on Saltspring Island as a cook, then subsequently worked in the Saltspring Island Creamery where he learned his trade of butter maker. This would be around 1900. There were many dairy farms on Saltspring Island and the other Gulf Islands. Saltspring Island butter was well known throughout Victoria and Vancouver and sold for a higher price. George Heinecke had his own dairy on Saltspring and another farmer in Vesuvius had a dairy herd as well. The milk would come to the creamery in 10 gallon cans.

Frank Norton Jr. 1994.

with Vancouver Island and made the move in spring of 1959. There were lots of farms for sale and it was a friend who finally introduced me to this farm where we have lived and worked since that date. The only reason this farm was for sale was this elderly couple had no family and he took ill and was forced to quit.

By April 1, 1959, we took over the farm but it was rather hard, having lived in a city for some years, to start dairying again. The herd consisted of Jerseys and Guernseys. The milk was poured in 10-gallon cans and on a little cart wheeled into the milk house. There, up to four cans were placed in a cooler where cold water was circulating around the cans to cool the milk as quickly as possible.

In the morning we had to milk early so the milk had time to be cooled before the cans were put in the pickup truck. At the corner of Crofton and Chemainus roads, we had to meet a CPR truck which would truck the milk to the Northwestern Creamery in Victoria, where it was bottled for the consumers. Sometimes the CPR truck was late and the milk would be on the road for too long. The farmer's concern was to deliver first quality milk, so any rise in the bacteria count could be disastrous.

We could expect a visit from the field man of the Northwestern Creamery with bad news if the milk was not standing up to the blue test, as it was called at that time. The result could be that the milk would be discarded and, worse, the licence suspended, as the flavour of the milk was of the utmost importance, equally what you fed the cows, as that can affect the milk. The field man from the dairy plant was very helpful.

Then the stainless steel bulk milk tanks were introduced in 1960. It was a big improvement to store and cool the milk, and we had it all in place by mid-1961. A wonderful improvement. For the next 10 years the milk was carried by pails to the milk tank but it was cooled right away. Some farmers used a "step saver" — with long plastic hoses — but cleaning them was a problem, and the bacteria count in the milk would rise.

From Switzerland to Fort St. John, 1950s
Otto and Ruth Wuthrich, 2000.

We arrived at Fort St. John in June 1949. We came from Switzerland through Saskatchewan, then came up here and worked for a man by the name of Bessey. He had bought a quarter of land and started milking cows and was looking for a helper to milk the cows. We milked cows for the summer — 30 cows by hand — and in November he told us there were no wages to be had in the wintertime, so we went into construction and worked at the airport. The following spring, 1950, he offered us the pasteurizing plant that was in an army warehouse at the airport. His son, Milton Bessey, wanted to quit the work of processing milk and delivering milk in towns. So Paul Odermatt and ourselves bought the plant and we processed milk. With the excess, we made Swiss cheese and delivered it to town. At that time, we had about 15 milk producers shipping milk in three- and five-gallon cans. In 1951, it became evident that we were going to run short of milk. We could see that our business could be jeopardized by the supply of milk. We had come over to Canada to farm,

Looking over the Jersey cow class at the Armstrong Fair, circa 1945.
LLOYD DUGGAN COLLECTION

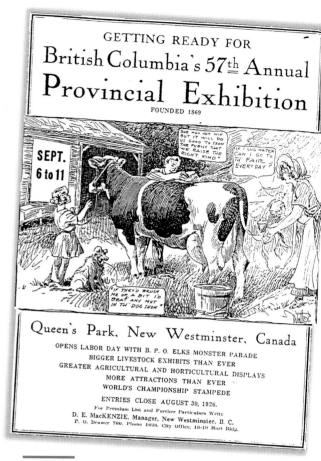

Advertisement for exhibition in New Westminster, 1926
BUTTER-FAT, AUGUST 1926

A Visitor to the Creamery

H.E. Church staked 320 acres at Big Creek, 20 miles from Hanceville, in the spring of 1903. In the summer of 1905, he sold $250 worth of butter locally. In the following years, He shipped butter by horse and stage from the Church Ranch to places in the Cariboo, especially mining camps. It sometimes took a week to get to its destination, but Church relied on stage drivers to do their best to keep the butter-boxes cool.

Quite early in the spring I got a tiny bear cub from an Indian who had killed the mother in her den. The cub was no larger than a smallish rabbit ... We had to keep "Buster" (as we called him) tied up a good deal of the time as he had an extraordinary knack for opening doors and windows. On one occasion he got into the dairy and indulged in a bath in the cream-vat, spoiling thirty gallons of cream ... He knew when the cream separator started running that it was time for his feed of skim milk and would come to the dairy door, stand on his hind legs, and when given a large milk-pan full of milk carry it a few feet away in his arms and then put it down on the ground without spilling a drop."

H. E. Church, An Emigrant in the Canadian Northwest, 1929.

so we bought this quarter section here, and Bessey offered his cows for sale. Odermatt kept on going with the dairy at the airport, and then eventually he built a plant in town in Fort St. John. He functioned with that dairy until 1963 and then sold out to Northern Alberta Dairy Pool. There was fierce competition between Odermatt and Northern Alberta Dairies.

From earlier times, there were some interesting stories. Only Mr. Bessey had hydro. Even here on the farm we produced until 1955 without power. It wasn't easy, the way we operated. We had to cool the milk by going to get ice blocks in the wintertime down at the Peace River and storing them in sawdust and then bringing them out and putting them into a water trough and putting the cans in the trough to cool the milk. Most farmers didn't bother with ice. They just cooled the milk outside, so you got all kinds of quality of milk. When I was at the dairy, with Odermatt, it was unreal the quality of milk you would see. I went to inspect a farm once because of the quality of milk, and they had no barn! The cattle were sleeping around the straw stack that they used to make with the threshing machines and that was the only shelter they had. For milking, he tied them to trees.

Another incident I remember was when I poured one of the five-gallon cans into the pasteurizing vat. We had a big sieve on top. I heard something strange go "gloop, gloop." I looked down and there was a sock, delivered with the milk. So I went to the farm to complain. I took him the sock. He was so excited that he had his sock back. Then he explained how it happened. At night he put his milk cans onto the porch. Over top

of the porch he always hung his coveralls and then put his socks on the milk cans. His daughter was told to go get some milk for the supper. In the dark, she just opened the lid and poured the milk out and the sock went in. It's something that I never forgot.

There was no quality control. We were a forgotten country up here. When we came, there was no connection to Prince George and the rest of the province. There was some good land to be had and we still had to clear the land to a great extent. There wasn't a decent animal to be seen as far as dairy cattle. There were a few Jerseys in the country. People brought them in for their cream. Back in 1957, we went out and bought our first purebred cattle. The first ones were from down by Grand Prairie — a farm with purebred Holsteins. We later got a truck load from Gilmore's sale and another truck load from Goldwood Dairies at Haney. Later on, in 1958, I trained at the BC Artificial Isemination Centre. Then I got busy chasing from farm to farm doing that job. Eventually, when the frozen semen came on stream, everybody received training and started looking after their own. Before the semen was frozen, we shipped semen by air twice a week from the BCAI centre. It worked okay, but you didn't have too much choice what sire you wanted.

Back in 1963, Odermatt sold to the Northern Alberta Dairy Pool and they took over the plant and then they built their new plant. In the wintertime there were big demands for milk because the oil industry had camps. They bought milk so much per head in camp, whether is was used or not: it was a good market for

Ruth Wuthrich with herd, early 1950s.

OTTO AND RUTH WUTHRICH COLLECTION

Clipping from the *Alaska Highway News*, 1968.

NORTH PEACE HISTORICAL SOCIETY

THE DAIRY HERD HAS ARRIVED

Fresh milk—real cow's milk—is now delivered daily in South Fort George. Last Monday a herd of a dozen milch cows arrived by steamer Chilcotin and are now being pastured at Six-mile Meadow. The lacteal fluid is retailed at two bits a quart and the demand is fully up to the supply.

The owners of the herd are two young men who recently arrived from the east. It had been their intention to run their herd on the Indian reserve meadow, where they had erected temporary buildings, but the recent order from the Grand Trunk Pacific, allowing the Indians the use of the fenced enclosures for this season, upset their plans for a location near town.

Clipping from the *Fort George Herald*, 18 May 1912.

PRINCE GEORGE PUBLIC LIBRARY

Belmont Farm

Our place was bush, heavy bush. Our dad came out of school in grade 10 to run the logging crew. He would hire a crew made up of East Indians and local men to clear with a block and tackle. That's what they set out to do and that's what they did. This place was all cleared by hand and all drained by hand. We spent a lot of our youth picking up roots.

Bill Berry and Chub Berry, 1999.

milk. So in the wintertime, the feeding system that was used in the Peace country was not up to par to what they do today as far as milk production. In the winter, milk production was usually down, and in the spring with the grass, spring you could sell your bull calves easy to the beef farmers. So there was all this surplus in the summertime and they really didn't have any use for it. They separated the milk and a lot of times you saw the skim milk go down the hole. That was a bad one. You had no stabilized market, so to speak.

With the Northern Alberta Dairy Pool, we established quota amongst us producers. We had an association that worked out its own quota in 1968. In bad months such as during spring break-up, or the month of May before the oil companies got going again, there was only 60 percent utilization. In the winter, there was 98 percent. When the Hudson's Hope dam got going, that was our best opportunity because milk was used year-round. They had a big camp with 5,000 workers. That was a lucrative time, and after that it just sunk right back down. We started negotiating with the Milk Board. They agreed that they would take us over in 1988 or 1989 and accepted the quotas we had established as a basis. It stabilized utilization so that you could know how many bucks you were going to end up with if you shipped your quota.

Afterwards, of course, Northern Alberta Dairy got married to Dairyland, and Dairyland got married to the Central Alberta Dairy Pool. That started changing the whole thing again, but the Marketing Board system was maintained so that the farmer wasn't the one who took the rap every time there was a change. With the

Milk Board, each farmer in the province is treated the same way.

The Northern Alberta Dairy Pool came into Dawson Creek in way back when in 1937. The Sudeten Germans were established when about 55 families that fled from Hitler came over. They came from Czechoslovakia and they developed an area south of Dawson Creek between the Alberta border and Dawson Creek. Most of them started shipping cream in cans into the Northern Alberta Dairy Pool in Dawson Creek. They had up to 50 shippers at one time and now there's one.

When we came into the country there was the Bessey Dairy, and years before, people told us there was different farmers that were peddling their milk in town. One of them was peddling milk in the hospital and someone got undulant fever. The government stepped in and declared that in the area between Dawson Creek and Fort St. John all milk had to be pasteurized milk. Bessey, who was one of the producers, put in a pasteurizing plant and another farmer, Whimpt, put in a pasteurizing plant. They were both at the airport. The airport was the only place where there was water available besides one hotel in Fort St. John. The army put in the pipeline from Charlie Lake to the airport. They let one hotel and a hospital hook on to the line. That goes back to 1942–43 when the Alaska highway was under construction. Bessey and Whimpt were in competition. Then Whimpt lost his bank roll and so he was finished. So Bessey's Dairy was operating until the spring of 1950 and then Odermatt and myself took over. After a year, I decided that I had come to Canada to farm, not

Chopping silage, Belmont Farm, Langley, home of the Berry Family, circa 1915.

Chasing Cows in Victoria

Bob Mercer was born down in the area of the Colquitz Junior High. He grew up on the farm there and then they bought land out on Blenkinsop Road. Bobby's old story was about when his dad had surplus cows and they were going to ship them off the Island, they'd drive them from the Colquitz area, down the road, down to the middle of Victoria onto the boat. No trucks, no wagons, no nothing. Just driving. And this would be the early 1900s.

John Pendray, 2000.

to peddle milk. Odermatt paid me out, and we concentrated on farming. Northern Alberta Dairy Pool took over Odermatt Dairy in 1963 and he went into a dry cleaning plant.

Twenty-eight years ago our daughter wanted to get married to a Swiss dairyman and we invited him to be part of the farm. Later on our son also wanted to join and we incorporated. They still run the dairy and we are the retired onlookers. I'm the "Go-For." I help sometimes — it is nice. Now we even have a grandchild involved, he was a helicopter pilot, he got fed up with being out in the bush camps and having his family at home growing up. Last October he joined the farm. He said, "I want my kids to grow up like I did." For me, it's gratifying to see the farm passed from generation to generation, to see something you've worked hard to establish continued on.

Farming in Greater Victoria

John Pendray, 2000.

I was born on a small dairy farm in 1925 in Swan Lake, inside the two-mile circle from City Hall. My dad moved to Victoria in 1910 and married my mother in 1916. They raised two daughters who became nurses, and two sons. My brother became head of the Grazing Division of the Forest Service.

Most milk distribution up until the 1950s was by producer-vendor. Our milk was delivered seven days a week to our retail and wholesale customers in the greater Victoria area, and we also distributed milk for my uncle's farm. The herds and farms were very small by today's standards, and it was mostly hand work un-

til the 1940s when milking machines and other equipment were introduced. Life was basically good. I guess we were poor, but we didn't know it because we were as well dressed as anybody else in the community. There were about 130 producer-vendors in the city of Victoria in the 1930s — and that would be from anybody with two or three cows to E. and T. Raper, the biggest one, with about 100 cows and two or three trucks on the road.

We were licensed and inspected. There was a dairy inspector in the city and he would stop you on the road and take a milk sample and do bacteria tests to check whether your milk was in good shape. When we were delivering milk, we were delivering what was called Preferred Raw Milk, which had to be below a certain bacteria count — in those days below 30,000. By today's standards, that's quite high, but by the standards of the day, that was very low. It was delivered raw and straight from the cow. We had our own good cooling system and the milk could be cooled down to about 34.5 degrees as soon as it left the cow, so that was one of the secrets to our success.

My brother, sisters, and I — like most farm kids — were fully involved on the farm. The work was divvied up among family members. The hard work didn't hurt us and we sure felt wanted! During the Second World War, farm help became extremely scarce, and the decision had to be made either to quit school or sell the delivery part of the business. I decided to continue in Grade 12, but delivered half a day's load of milk before I went to school. I was delivering 125 to 150 quarts. That meant that I was up at 4:00 a.m. which was

John Pendray, 1990s.

Good Judgement

A life-long dairy farmer, John Pendray also served as president of the board of directors of Island Farms for 29 years. John possesses a high level of business acumen that extends well beyond the farm, and he applied it vigorously to the success of the co-op. He is a natural leader and was relied on by the farmer-owners for his good judgement when difficult decisions needed to be made. His career-long goal was to ensure that Vancouver Island farmers kept control of the marketing of their milk, and the substantial success and independence of Island Farms attest to his efforts.

Des Thompson, 2000.

hard, but I don't think it hurt me too much. That was in 1942–43 and I continued in this manner for two years of university. At that point, I decided to stay on the farm full time instead of going to UBC, which maybe was a mistake, but I don't really regret it.

The milk routes were sold in 1951 to a private dairy and we concentrated on farming and milk production. But shipping to a private dairy didn't really work out so well, so in 1961, we moved to Island Farms and we've been very happy about the move ever since. It's interesting to know that some time in the 1950s, there were about 70 dairy farms still on the lower Island — from Sooke to Sidney. Today we only have five. Mind you, the farms are all much bigger and shipping a great deal more milk, but certainly the numbers have moved.

My wife, Joyce, and I raised four children. My daughters have moved onto other careers after working on the farm while they were students. The herd has been gradually expanded. Our son, David, started working on the farm in the late 1960s and our son, Michael, in the 1970s. Today my sons run the farm and basically own the operation. In the 1940s, we started Dairy Herd Improvement. We also started using AI in the 1940s from the local AI club and later using BCAI sires when they came on stream. This combination of tracking an individual cow's production, using proven and selected sires, and improved feeding has greatly increased production per cow throughout the dairy industry.

In the 1970s, it became apparent that our farm on Blenkinsop Road was too small and the rented land was being committed for other uses. So our present land

in Saanich was purchased in 1978. It had not been used as a dairy farm up until that time and a complete set of new dairy buildings had to be put in. Drainage and irrigation work had to be done. Although it was a very expensive situation, it gave us a modern and efficient operation.

The top producing herd of 50 years ago would be about 5,000 to 6,000 kilograms and most people were much smaller than that. Today, many herds are averaging over 11,000 kilograms — not uncommon at all. Many tasks that formally took a great deal of hand labour are now accomplished by having improved buildings, milking machines, pipe lines, bulk tanks, and milking parlours. Cleaning barns by hand is basically a thing of the past. Mechanization has improved productivity and actually makes the work much more pleasant in a lot of ways. People hired on farms today are people in charge of many hundreds of thousands of dollars worth of cattle and equipment. At one time, you used to give employees a bucket, a stool, a broom, and a shovel and that was your investment. The kind of people we need to have on a farm today are quite capable people that could take a job pretty much anywhere in the city. The technology and business knowledge of the farmer has had to keep expanding and improving. Farmers have to be able to innovate and be on top of so many different areas of expertise today — from computers to veterinary science to soil science and good business practices. But having said that, dairy farming is still a way of life. The most important thing is to make maximum use of the resources available.

It is difficult to keep everything in balance while

NOCA dairy products float, Vernon, likely the 1960s.

The CBC Farm Broadcast

I can't imagine a history of any farming in BC without mention of Tom Leach. Remember when we were all young and the CBC Farm Broadcast came on at noon. Everyone (almost) tuned in to get the latest on farming and a trip to the family at Willow Brook Farm.

Dr. A. Kidd, 2000.

working in any family business. All the partners know each other very well, which at times can be awkward. In the case of our family farm, we have always tried to keep things modern and up-to-date. That always appealed to me and, of course, that appeals to the younger generation. They liked the lifestyle, but they also liked the fact that if there was a better way of doing things, we usually tried to do it. If we could increase productivity or eliminate some of the hand work, that's what was done. I think that's why you'll find that there are still many young people on BC farms. Letting the younger generation make decisions on their own encourages them to stay with the family farm. Some people wouldn't like the farming lifestyle, but others love it. The existing farms in our area are all family operations. Generally, it's worked out very well here. The operation is successful. I've always felt that I've been a fortunate person to be born where I was, when I was, and to be given the opportunities that I've had — both to be a dairy farmer and to have been associated with Island Farms for many years.

In my role as a board member of Island Farms, my main goal was always to ensure there was a viable market for our products. When I became president, there were three dairies in Victoria and Island Farms was probably second in size. Thanks to a supportive membership, a forward-looking Board, and a good management team, including Fred Mockford, Des Thompson, Reg Owens and others, Island Farms grew and we were able to purchase Silverwood Dairy. At that time, we took on a large loan to expand and revamp the plant. With hard work we soon paid off the debt, which was a relief for me having had no prior experience with loans of that magnitude. But it worked out even better than we had projected. In 1989, Palm Dairies came up for sale and we purchased its business on Vancouver Island. Now all producers from the Parksville/Qualicum area to the Saanich peninsula ship to Island farms. The organization is efficient and the farmers are well represented on the Board by six directors. We have two general meetings a year to ensure the farmers get a good picture of what's happening in the business they own. When we were making big changes, we held membership meetings in the plant to show members through, and to show them how the money had been spent. It sure did away with the question, "Where has all the money gone?" We always got great cooperation — we really did. I felt that without communication and cooperation, dairy farms would disappear from the Island.

Farming has great economic influences on a community. Economists say that the money generated on a farm is circulated five to seven times. One farm could have about a $10 million influence on a community. Think of all the other dairy farms and also the economic influence of the approximately 200 employees of Island Farms.

Dairy farms are important for a lot of reasons besides economic spin-offs. As the metropolitan areas fill in, the open green space is worth a lot to the people living in the city. It is generally well cared for and is essentially free, open space to the taxpayer. An area like Victoria is getting crowded. I think that as time goes on, as long as farms like this can exist, they will be

Comox Valley Calf Club, 1950 or 1951.
JIM CASANAVE COLLECTION

Tug-of-war, Cow Testing Association, Matsqui, 1939.
JAKE SAVAGE COLLECTION

Jack Worrell of the Langley Calf Club, 1935.
JACK WORRELL COLLECTION

213

providing a greater and greater good to the general community as open space. Whether this particular farm will exist into the future is just anybody's guess because in today's market, this farm would be worth more broken up than it would sold as a unit. But having said that, I hope it doesn't happen and I don't think it will in a hurry.

Today, the farm I was born on is a park — it seems like it's in the middle of Victoria. I look at it and I think of how people like my dad got started and of his fortitude in doing it. He was the youngest of six boys on a farm in England. When he was only six, his dad was killed in a farm accident. Eventually two of his older brothers came out to Ontario, and they liked it out there, so the rest of the family — four other boys and the mother — were going to come out. They all went down to Southampton. When they got there, because one of Dad's brothers had polio and there was nobody financially capable of looking after them, they weren't allowed to come. Right there on the dock, Dad, who was 14 and his brother John ,who was 16, said goodbye to their brothers and their mother and got on the boat. Dad never got home to see his mother again. His pioneering efforts in our dairy industry gave myself and our two sons the opportunity to be dairy farmers. I'm proud of what we, as a family, have accomplished.

Changes in the Crescent Valley

Ray Kosiancic, 2000.

I have lived all my life on the farm that I own, as did my father. The farm was pioneered by my grandfather at the turn of the century. The farm meant a lot to me:

when my uncle decided to sell, I said I would like to buy it. He laughed and wondered how I could even raise the down payment. Most farming neighbours told me I was crazy to go into dairy farming, as many wanted to get out of the business. If I had to do it all over, there would be little change, as I operated very thriftily with good planning and lots of hard work and long hours.

For my family and the farm that Ida and I own at Crescent Valley, the last century has been an important one. In 1898, my grandfather, Jacob Kosiancic, and his wife, Antonia, left Triest, Austria, and boarded a steamship for North America. He worked in a steel mill in Pittsburgh for ten cents an hour, twelve hours a day, six days a week, forging railway steel. This is where my father, Valentine, was born. Hearing of the gold rush in British Columbia, Jacob and his family travelled to Rossland in 1900 and worked in the Leroi mines for $2.50 a day. At this time, he applied for 400 acres of crown land here at Crescent Valley and paid $2.50 an acre. Whether it was by accident or good fortune, this lovely piece of land that overlooks the Slocan river is now the farm that we own.

My grandparents had two more sons, Joe and Jack. Over the years, the farm became known as Kosiancic Brothers. They lived in a log and shake building with dirt floors that was constructed during the building of the CPR in 1893 and was used as a bunkhouse. Fish were plentiful and so was wildlife. My grandmother would go into the forest with a double-barrelled shot gun in the afternoon and shoot a rabbit for supper.

There were only six families between Crescent Valley

Kosiancic Brothers Farm, Crescent Valley, 1945.

Jacob Kosiancic.

Jack Kosiancic with 1928 Dodge sedan used to deliver milk, 1930s.

Hoping For Cows Near Invermere

This morning we were up with the lark and had had breakfast and made the bed and swept the barn and had the breakfast things washed and the boots cleaned and started for Invermere at 8:30. It was a lovely morning with blue sky and a soft light breeze, and everything was looking its best. First of all, we went to the "Experimental Farm" and interviewed one Mr. Anderson by name, who is in charge, as to why our potatoes were not doing their best ... We shall have enough potatoes for ourselves but we shall not make a fortune this year by selling them. Of course, not having a horse at present Jack could not keep the soil "going" as well as other people who have a team ... But we are not down-hearted yet and look for better things next year when we shall have a horse and perhaps a cow. Your dairy knowledge will come in useful!

from Daisy Phillips to her sister, Freda. Letter dated Aug 7, 1912. In Letters from Windemere 1912-1914, edited by Cole Harris and Elizabeth Phillips, 1984

and Slocan at the turn of the century. There were no vehicles and only a wagon trail in the Slocan Valley. The CPR steam train would bring in supplies for the pioneers and prospectors and then haul out railroad ties, lumber, and mined ore. Jacob Kosiancic applied to the railroad company for a siding in 1905. It was approved providing he moved hundreds and hundreds of yards of earth for a proper grade. He worked all year from daylight to dark shovelling most of the soil by hand. Finally, he had his own siding called the Kosiancic Spur. He loaded cordwood into railcars and shipped it to Consolidated Mining and Smelting, now known as Cominco, for $2.50 per cord. He raised his own team of oxen, and like all pioneers, his family raised cattle, pigs, chickens, and grew most of the vegetables that were used in their house.

He cleared land in the winter: huge trees were cut down and skidded with a team of oxen or horses. A large pole barn was built in 1909 with a shake roof and shake siding. In 1911, Grandad built a large two-storey farm house of hewn cedar logs. It is still on the farm today. My grandmother was very religious and one of the many rooms in the farm home was like a little church. It had pews, religious pictures, statues, and stations of the cross. Priests would often come out to say mass and nuns would also come to the farm as a retreat.

Apple trees were planted in 1913 — 250 of them. By 1925, 700 boxes were loaded into two CPR box cars. The shipment went to Associated Growers in Nelson for $140 — barely enough to pay for the boxes and the packing. A 50-cow dairy barn was built in

1914. Milk was shipped by train to the City Dairy in Nelson. There was no train on Sunday, so milk was taken by horse and wagon half way to Nelson where it was met by a teamster and his rig. Milk was also delivered to neighbours on horse back in a five-gallon can and ladled out for 10 cents a quart. All milking was done by hand until 1916 when the first milking machine in the Kootenays was purchased. It was a vacuum pump driven by a single cylinder stationary gas engine.

My uncle Joe took a course in cheese making in Vancouver in 1920. Approximately a thousand pounds of cheese was made each month and sold mostly in Trail. Jack Kosiancic drove a 1928 Dodge sedan to deliver milk in the area and to haul cheese to Trail. A market garden was started in 1931, and a Chinese worker, Der Yew (Dee), was employed to grow all vegetables and some fruit. We always prayed for a good harvest, and we always kept hoeing.

A large, three-storey chicken house was built in 1933 to hold a thousand hens. Fresh eggs and vegetables were sold on the milk route. Entering the Depression, or the Hungry Thirties, farm wages were $1.00 a day, and potatoes sold for $1.50 per hundred pounds. The dairy was the mainstay of the farm, along with root crops. In 1936, a new International milk truck was purchased, and in 1937, my father, Val, bought a new three-ton White truck with a 14-foot steel flat deck and a hydraulic dump to haul logs, lumber, potatoes, hay, and also farm produce. Some three hundred tons of netted gem potatoes and a hundred tons of K.B. turnips were grown each year.

Dairy farm, Rossland, 1945.

Minding Children while Milking

In the evening there were cows to be milked, milk to be separated, and calves to be tended. With the hot stove and oil lamps, Mother was afraid of fire and could not leave me alone in the house, so while she was milking I was tied to a pole in the barn! Our cream was sent to the Creamery in big cans. Milk was used to fatten our pigs and calves. The Creamery would empty the cans and send them back unwashed. There was always some thick cream left in the cans and this would be sour from the heat of the day. It was too good to waste and Mother used it in cooking, particularly to make some buns that she called ragged jacks.

Mary Elizabeth Flett,
from Memories Never Lost, *1986*

In 1955, I purchased the farm with my wife, Ida, and continued the dairy business, known as Raida's Dairy. My interest in vehicles, a trait I inherited from my dad, meant that I delivered milk with a custom 1964 GM pickup with a 353 Detroit diesel engine. This special conversion was the only one in Canada. Most of my customers remember hearing the screaming Jimmy on the milk route. The dairy business was discontinued in 1972, but we continued farming, raising beef cattle and hay until 1995 when all the cattle were sold. Let us not forget that the cultivation of the earth is one of the most important labours of man. I do not miss the old manure shovel, the pitch fork, the milk cans, the dusty jute feed sacks, the 45-gallon gasoline and fuel barrels, or the heavy steel irrigation pipes and sprinklers, but they are part of the history of the Kosiancic farm.

The Comox Valley

Jim Casanave, 2000.

My grandfather was a Basque and came here in 1860 as a single man and bought, I believe, 40 acres in the Oak Bay area. I'm not sure when he was married — he married a woman from Los Angeles and there were seven children. My dad was the fifth out of the seven and was born in 1891. He quit school in Grade 8 — as most people did then — and took over the farm at home because his father really wasn't a farmer, and he milked cows and delivered milk on a milk route in Victoria. The farm was on the site of Oak Bay High School.

At some point around that time, grandfather bought 28 acres at Gordon Head. In 1912, my dad built a barn out there, right near the university, and then he moved his milk cows and he started farming there until the war. When the war started, he, of course, went to war like every other young fellow and was in Europe till the war was over. After the war, he never farmed in Victoria again, but worked in a sawmill in Vancouver and then got a job on a farm in Prince Rupert for two years and then went to Courtenay in 1921. Now this Courtenay farm was bought by my grandfather also as a speculative venture about 1912 or 1913. After the war, my father came from Prince Rupert back to the farm in Courtenay, of course where he's been ever since. Lucky him, he had me in 1941.

When he left for the war, he left his team of horses and his few other odds and ends with a friend of his. He said, "If I don't come back from the war, they're yours, keep them." So he came back from the war, so the guy just kept the horses until dad wanted them and so on, and he ended up with a team of horses, a cow, a calf, six hens, and a rooster and a milk cart. That's what he brought from Victoria. That's what he started with in Courtenay. He was still single in 1921 and he attempted farming or did the best he could. He was married in 1927, but that's how he got started and it was mixed farming, of course — everything was at that time. My father was a farmer, period. That's it, a horse farmer. The first tractor was bought on that farm in Courtenay in 1947, so everything was horses from 1921 to 1947.

The Comox Creamery was the only place you could ship milk to. I think it was there at the turn of the

Marygrace Casanave, Courtney, 1948.
JIM CASANAVE COLLECTION

Left to right: Jake Born, Bill Wiebe, and Jake Tilitzky hauling vita grass in Sumas while earning money to pay for dairy farm at Mt. Lehman. Hilda Born farmed during the day. 1953.
HILDA BORN COLLECTION

Victor Casanave (left) and Jim Casanave Sr. (right), 1912.
JIM CASANAVE COLLECTION

Paulholm Holsteins

When Mr. Paull came to the Fraser Valley in 1925, he had a carload of settlers' effects, including 11 cattle and a team of horses as his worldly goods – and a heavy debt. His family deserves equal credit with him for the success that has been made. His three sons are partners with him, Elmer being herdsman and quite as much a Holstein enthusiast as his father.

Hugh Colson, "Notes from a Western Trip," The Holstein-Friesian Journal, September 1939.

century. He shipped to Comox Creamery, but we grew potatoes and poultry and you name it — like everyone else. There were milk routes in Courtenay, but we shipped directly to the creamery all the time. Originally you had to separate your milk, you only shipped the cream, and, of course, fed the hogs. We had a separator at the farm. I know, it was still there when I was a kid. But fluid milk was after the war, during the war and after the war, when everybody started shipping fluid. There was a lot of cream shippers before that time.

As children in Victoria — my dad having been born in 1891 — they pastured their cows hither and yon, but one of the main places was what's now known as the Uplands. The children would take the cows to pasture in the morning and go get them after school and bring them back in from the Uplands. Building up the farm was a very, very slow process. In 1972, I bought dad out. At that point, he had 3,000 pounds of quota, which would equate to about 1,300 litres. That would be what I bought from him. My grandfather, he did gravitate to heaven — which is, of course, Vancouver Island. That's probably why we're all here. That's where I made my fortune anyway, milking cows. I followed this man's footsteps.

The Story of Frasea Farm
Jake Grauer, 1973.

I was born on the family farm on Sea Island on February 23, 1902, the seventh child in a family of nine. I was christened John Jacob Grauer, my father's namesake, and nicknamed Jake. Father and Mother first met in Seattle, Washington, where they were married in 1885, and almost immediately migrated to British Columbia. He leased a farm on the north-east part of Sea Island in 1890 on which he grazed and fattened livestock to market through his butcher shops. There was no bridge across the Fraser River connecting Sea Island and the mainland at that time. Access to the Island was by boat and scow only; however there was a cable across the river from our farm to the north shore approximately at the end of what is now Hudson Street. A scow large enough to carry a team of horses and wagon was attached to cables and by a mechanical device you could manually crank or wind your outfit across the river.

Soon a bridge was built to replace the cable setup at the same location, and my father's headquarters were permanently established on the farm on Sea Island. Following the Eburne bridge, a bridge was also built across the Middle Arm or Moray Channel of the Fraser River connecting Sea Island and Lulu Island. Father purchased the farm at Eburne about 1890. Over a period of years he purchased additional farms on Sea Island and also a section of farmland bordering Boundary Bay in the Delta Municipality.

My mother was a pioneer in the fullest meaning of the word. Her loving disposition and tremendous energy left a lasting influence on her family of seven sons and two daughters, also the church, and the community. She retired and lived in Vancouver with her daughters where she passed at the grand age of 96 years in the year 1963. Her whole married life of 77 years were spent on the farm and in Vancouver.

I can remember the North Arm of the Fraser freez-

Employees, Frasea Farms, circa 1940.

Farming in Richmond

I have lived here all my life — 84 years. A lot of things have happened in my time here. One dairy, Fraser Valley Milk Producers' used to have about 60 or 70 shippers in Richmond. Today, there are three farms left shipping milk. We got tired of shipping milk for a dollar a can, so we went to the bank, borrowed some money, and put in a boiler so we could wash bottles and bottle our milk on the farm. In those days, you could get a permit from the city to sell fresh, raw milk. Jim Holt had a dairy, Savage Farms had a dairy, Mitchells had a dairy. Then Richmond Dairy wanted a source of raw milk, so they came to us. We bottled for Richmond Dairy for quite a number of years, sold the milk under the Savage Dairies label, and shipped any surplus to them.

Doug Savage, 2000.

ing over solid from Iona Island at the mouth of the Fraser River to New Westminster. This happened on several occasions. Hundreds of people skated on the river. One year in particular the horse and wagon traffic used the river ice to travel on in preference to the deep drifted, snowy roads. Also at times the bridges were declared unsafe and closed to traffic as a result of the ice jamming against the bridge piling.

Farmers on Sea Island and Lulu Island were obliged to build dykes along the foreshore of their farms to avoid flooding from the Fraser River, particularly during the great freshet or spring runoff when warm weather melted the snow in the mountain areas. At an early age I recall arising in the morning amazed to see our whole farm under about four feet of water. The dykes had broken and water from the turbulent Fraser rushed in and inundated the whole Island. Similar flooding occurred before and in subsequent years.

During this particularly flood around 1908, a substantial number of our sheep were drowned. Hundreds more would have shared a similar fate had not a fleet of fishermen in their boats rowed across our pasture fields and rescued the remaining animals. Flooding of Sea Island was eliminated when the Richmond Municipal Administration finally had a permanent dyke built around the whole island. In modern times pumps were installed to assist in draining the land. Our farms were all under-drained with miles of drains made from cedar lumber which emptied into open ditches which in turn drained into the Fraser River. Those were all dug by hand with spade and shovel.

At one time my plans were to continue my studies in technology and become an electrical engineer. However my father, with his tremendous faith in the land, finally persuaded me to attend Agriculture College and become a farmer. The love of animals and the farms were indeed close to my heart and I enjoyed my boyhood days on the farm beyond words. However my decision to pursue a career as a farmer emanated from a much deeper reason than simply my own sentiments: I was very grateful to my parents for the love and care I received from them during my boyhood years. I wanted to indicate my gratitude to them every way possible. They were Christians and by word and deed they instilled in me during my youth the Christian principles that meant so much to me as the years rolled by and problems and responsibilities became more pronounced. Consequently, at that time of decision, I felt that I should accede to my father's wishes and assist him in expanding his agricultural enterprises on a more scientific basis as soon as I could become involved. My schooling was completed at the Oregon State Agricultural College and I returned to the Sea Island farm in 1920 and went to work. At that time my two oldest brothers, George and Gus, were operating the farms.

A foundation herd of purebred Holstein-Friesian cattle were obtained and the first dairy unit was established by building a 50-cow barn in 1922. In 1927, my father formed our family company and thereafter our entire farming operations were consolidated under the name of J. Grauer & Sons Ltd. Each member of the family held shares in the company with the exception of my brother, R.M. Grauer, who was firmly established in his own grocery and meat business and did

Frasea Farms complex, circa 1940.

Producer-Vendors in Richmond

I was born in Richmond in 1922. I grew up on the farm. There were at least 25 or 30 people delivering milk in Vancouver at that time, from little farms with 10, 20, or 30 cows. That's what we started doing — bottling milk and delivering it to houses. My wife and I were delivering milk after we did all the rest of our work during the day. When the war came on and we couldn't hire people to work, we sold out to Grauer's Frasea Farms on Sea Island and carried on with our farming. My father started with mixed cows, mixed grades. Then we started peddling milk and people wanted a little more cream on their bottle, so we got a few Jerseys and then we started buying better bulls from Jake Grauer. They all turned out to be very good for us – they improved our herd because they were registered Holsteins. Then the kids came along and we got them into 4-H. Then they wanted some better calves, so we got some calves with a little bit of show and style to them. It improved our herd.

Jim Holt Sr., 2000.

not with to join the family operation. Four brothers and two sisters became actively engaged in the company's operations headed by father as president. The two youngest brothers began professional careers: Albert, or Dal, eventually became president of BC Electric; Fred practiced surgery in Vancouver after serving in the Second World War.

The establishing of a dairy herd presented a milk marketing problem. Not being successful in negotiating a satisfactory contract to sell our milk to the then existing milk distributors in Vancouver, we decided to market our milk direct to the consumer from our farms. Initially, we sold milk in its natural raw form as the City of Vancouver ordinances did not permit milk to be sold from pasteurizing plants beyond the city limits.

To meet the demand for a rich creamy milk we established a Jersey herd of cattle in addition to the Holstein herd. Only the highest quality animals for both type and milk production were selected. Over a period of years a considerable number of foundation stock were imported direct from the Isle of Jersey in the English Channel where the Jersey breed originated.

Largely through natural increase and a few select purchases the two herds increased to approximately 300 Holsteins and 200 Jerseys. The herd was limited to approximately 500 head until it finally dispersed in 1954. Surplus stock was sold until breeding purposes and gave us a very rewarding income.

Prior to 1925, the major crops grown were hay, oats, wheat, barley and a small acreage of potatoes and mangles, mostly sold as cash crops. Prior to around 1925 all the farm work was done with horse-drawn

equipment. Some sheep, swine, and beef cattle were raised. We bred and raised many sheep, swine, and horses, including riding horses. The crops grown were suitable for dairy cattle and milk production. Large acreages of clover were grown for the silos. Green clover was also cut daily and fed to the animals in preference to pasturing the fields, as we found this a more economical method.

Legume hay, victory oats, silage, and mangles were all fed to the dairy cattle and marketed in the form of milk. With the fertility of the land restored, additional crops such as potatoes, peas, and red clover seed were sold as cash crops. Peas were sold to the Canadian Canneries in Vancouver for canning and freezing.

In 1936, our beloved father passed away at the age of 76 years. He was the first casualty in our family group and his death was indeed difficult to accept. The year 1938 proved to be quite eventful in the progress of our business. We equipped a very modern milk processing plant on the farm and we were in the pasteurizing milk business with great enthusiasm. A subsidiary company was formed to separate the milk processing and distributing business from the farming operations as they were two entirely different types of businesses. The subsidiary was named Frasea Farms Ltd. Frasea was also used as a prefix in naming all of our purebred stock.

Our dairy was unique compared to the dairies in operation at the time and with whom we were to compete. Owning our own herds and processing our own milk through our own plant straight from the farm direct to the consumer's home by our own delivery units

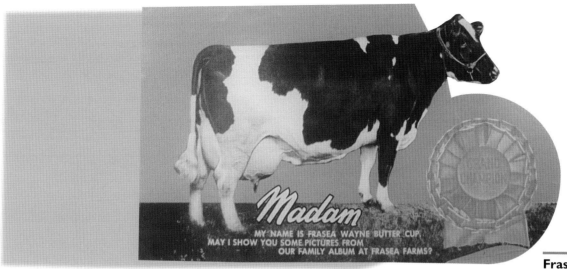

Frasea Farms brochure, circa 1950.
BCARS GRAUER PAPERS

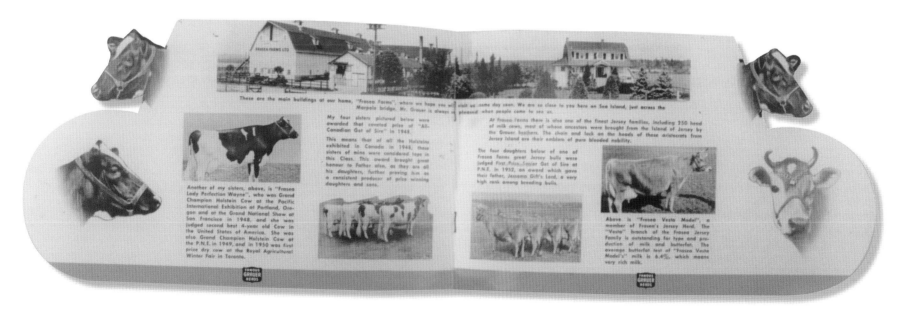

225

The Greatest Herd I Ever Saw

Of all the many places I wished that I could stay longer it was even more so at Colony Farm looking at the greatest herd of cows I ever saw – all home bred – and listening to that master breeder Pete Moore explain more sense about Holstein breeding than I ever heard before ... As you walk down the barn you say to yourself, "What a cow. She must be the best on in the herd" – and then in a moment you see another that you think is better. So it goes as long as you look.

Hugh Colson, "Notes from a Western Trip," The Holstein-Friesian Journal, September 1939.

was not only ideal, but very attractive from a sales and advertisement point of view. The proximity of our farms and dairy to the market was of tremendous public relations and advertising value. Consequently, we encouraged school classes to visit the farm. They were treated to dairy products and cookies after being escorted through the dairy and barns. We had numerous visits from customers, women's organizations, public health and nursing groups, stock judging classes — local and from the State of Washington. On one occasion we hosted the Vancouver Junior Board of Trade for luncheon in the loft of one of our barns. Our motto was "visitors always welcome."

The greatest milestone in my life had to be 1938. I married a lassie from Glasgow, Scotland — Margaret Downie. We went to California on our honeymoon. This was the first real holiday I had since 1920. War clouds were beginning to form now and in 1939 our country was faced with World War II. By now our milk sales were increasing nicely and we were obliged to augment our milk supply by purchasing milk from other dairy farmers in Richmond. We churned and sold butter and also cottage cheese. The scarcity of farm help during the war was very serious. Tires and gas were rationed severely. Finally, new trucks and used trucks for milk delivery were simply not obtainable. It occurred to us that perhaps milk delivery by horse and wagon may prove feasible and thus relieve the milk delivery quandary. We had oats and hay but no gas. Three wagon routes were started on a trial basis and proved practical. By the end of the world war we had 13 horse-drawn wagon routes in operation. Our number one problem was to find men who could drive a horse.

In 1954, most our farm on Sea Island was expropriated by the Government of Canada to be included in the airport expansion. It was at this juncture that the members of the family company decided to sell the entire assets of the company including land, herds, and milk business. The sale was accomplished in that year. So terminated the saga of one of Richmond's pioneer family farm enterprises, which participated in the early history and development of the Municipality of Richmond for almost 70 years.

Pioneering in Cobble Hill
Bill Wilkinson, 1999.

I was born in Victoria in 1917. When the war was over, my father worked in Esquimalt as an instructor. In 1919, we moved down here and rented the house at the road's end. My mother decided where we were going to put the house. She came down here. She could hear the creek, but she couldn't get to it. There were big logs along the creek which had fallen. The idea was they were going to move these logs in the creek and pull them down to the estuary where there was a mill. A lot of them didn't get down, they just rotted in the bush. Anyway, she could hear this water and she said, this is where we'll put the house. And the house was this kitchen. That was in 1919. The family grew quite rapidly. I was born in 1917 and my youngest brother was born in 1930.

There was nothing here except, I think, a small cottage up on the hill. It was mixed farming. We shipped

FVMPA shippers have a word with **FVMPA** president Danny Nicholson, early 1950s.

Looking for Tokens

We were looking for milk tokens from Registered Jersey Farms in Saanichton and were referred to the home of Bert Doney, one of the original owners. The night we arrived at the farm, we found that the Doneys had moved ... Finding the new home, we were told Bert was over in the fields fixing the tractor ... For about an hour we tried to help Bert and his son to get the tractor started. Bert was too preoccupied to even think about the old tokens, darkness was growing, and it looked like the trip was to be wasted. Eventually, though ... Bert took us up to his old farm. He thought that he had buried the tokens near the old outhouse some 10 or 20 years before. He pointed to two big trees, saying that the tokens were probably at the base of one or the other, then he bent down and scratched at the surface covering of fir needles and grass and picked up an old token!

Ron Greene, "Registered Jersey Farms," 1967.

our cream by train from the Copper Hill station. Copper Hill was a booming river village in those days. They had everything there. It went to Northwest Creamery in Victoria.

I don't know the exact dates, but I know we milked by hand for years. My mother and Dad used to do the milking. They used to go down after supper and us kids were left in the house to study and do our homework. It was so long ago. I still have an old gramophone upstairs that Dad picked up at an auction. It played records and some of the old-time music. I haven't played it in over three years, but it brings back a lot of memories.

We shipped the cream once a week. One time, there was a can of cream sitting on the end of the veranda and I decided I wanted to move the can. So I picked it up by the handle. The handle came off and down went the cream. Down the steps and everything. I was in deep trouble. My father wasn't very happy about that. It was more than the week's groceries going down the steps. What I remember about those times was that nobody had anything and you made your own fun. And we had a lot of fun. I can tell you this much, there was nine of us and I can't remember once getting into a fight with my brothers and sisters. I'm not going to say this is the usual, but I am saying I just can't remember any fights.

My parents were milking around seven or eight cows, something like that. And they had chickens. When the cows weren't paying, the chickens were, and when the chickens weren't paying, the cows were. That's the way it was in those days. Life was very isolated. You never travelled far; in fact, my dad got his first motor vehicle in 1927. I would be nine or ten then. It was a

Model T, a half-ton pickup. I don't remember going over the Malahat that early, but people did go over in Model T's. There used to be a spring coming over the hill with a tap on it. Before they went up, trucks pulled in there and watered up. We used to go to Victoria probably once a year. There was only one school in the Mill Bay area. And the Cobble Hill School. But they didn't have a high school, so when I got through public school, I worked for my father for five dollars a month. Everybody was itching to leave home, and that was a big move. I can appreciate my father — he was a great man, he had great principles, and I appreciate him today. I went down to the barn one morning and said, "Dad, what is there to do this morning?" He said, "You can do whatever you want." Well, I did. I had a Model T truck then, and I used to be able to work with that, so I left home. I stayed with a neighbour for over a year and one or two neighbours tried to get me home again, but I didn't go. I got a job on a farm at Hillbank for two or three months, $25 a month. And then I got another job with logging the yellow cedar up at Double Bay across from Port Hardy. That didn't turn out too well. I didn't mind, I was young. I was mainly getting experience more than anything. That was more value to me than money. So anyway, when I got up there, there was no yellow cedar to be logged, so that didn't last very long. I came back and came home here for a little while and then I went back to work again. From there, I just picked up odd bits of work around Maple Bay. Even cut the wood by hand, I used to cut my cord of wood, stove wood, and take it into town and sell it for four or five dollars.

Sookeway Dairy, the home of Mr. and Mrs. John Dorar, 1914.

When war broke out, we signed up right at the start. You got your clothes and your meals and your board thrown in all for $1.25 a day. It was all experience, wonderful experience. I don't regret it one bit. I spent six years in the war. One day, Dad rode over and said, "Are you interested in the farm?" That was quite a surprise because I had no intentions of going farming. I came home, put my books away, and never looked at them from that day to this. I says, I'm just going to work hard and get caught up in my work around here and then take a day or two off a week, but that never comes! The first year I worked the farm with the old horses, then one day I was plowing and it was slow going. A neighbour came down with a tractor, and he had this double furrow plow on the back. I said, "Do you mind if you let me try this plow out on the tractor?" Well, I got the whole work done in five minutes. So I bought myself a Ford and two or three pieces of machinery and I went out and did some custom work. When I first went out, I earned $2.50 an hour for the tractor and myself. So that's how things were in those days. Initially, it was mixed farming. I had everything growing here until I found out I had more than I could handle, and then I cut back and specialized in dairy and I still had pigs. I built a cottage up there on a shoestring. This house up top had to be finished before my wife and myself moved back to the farm. We got it finished and I asked my wife, "Which house do you want?" She said, "The one on the farm." I'm glad she did, too. It holds many memories, this house. But all this wasn't done without a lot of hard work. When I took over the farm, Dad advised me to ship to Island Farms. He wasn't; he was still shipping to Northwest Creamery then. So, I had to buy two cans to be able to ship to them. Two-hundred dollars a can to ship to Island Farms! That was your investment.

You have to like farming to carry on. It is a way of life; it's a routine and if you have to work seven days a week, you work seven days a week because you have to make sure you're cared for. Anyway, today I've got a farm with everything paid for, and I don't want to sell it. I'd just rather turn it over to the kids.

Farming in Chilliwack and Dewdney, from 1900 to the 1950s

Harry Irwin, 2000.

I was born on a farm right on the Lickman Road in Chilliwack on the seventeenth of September 1899. My dad used to have a lot of heavy horses, sheep, and cattle. But, in those days, the only creamery was the Eden Bank Creamery. Farmers used to take their cream there. When the Creamery was moved more into Sardis, the farmers had a store in connection with it. They made butter and cheese and you could go in there and haul the skim milk out back for your pigs or your calves. Everything had to count. The Chilliwack Creamery was at the south end of Young Road. It was quite a big creamery and was cooperative. A lot of milk was shipped there and made into butter and cheese. The only transportation was by boat on the river. The dairy never could come into any profits until Sumas Lake was reclaimed. The mosquitos were terrible — in June, July, and August, your production went straight down to nothing, so draining the lake was the greatest thing that ever happened. Minister of Agri-

Takes Away the Drudgery

NOT only to the farmer has the introduction of electric power proved a wonderful boon.

The housewife, who until recently did all the sweeping, washing and similar manual tasks including the churning, now leaves all this to electricity.

She is free to give her time and her energy to more important work around the house while electricity takes many of these routine tasks off her hands.

Our nearest agent will explain how you can get electricity.

BRITISH COLUMBIA ⬥ ELECTRIC RAILWAY CO.

HEAD OFFICE   VANCOUVER, B.C.

F15

FARM WIVES Are SOLDIERS *Too!*

CONSERVE VITAL TIME AND ENERGY USED IN IRONING FOR OTHER IMPORTANT FARM

WAR EFFORTS

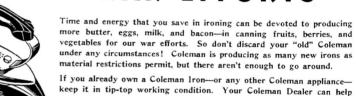

Time and energy that you save in ironing can be devoted to producing more butter, eggs, milk, and bacon—in canning fruits, berries, and vegetables for our war efforts. So don't discard your "old" Coleman under any circumstances! Coleman is producing as many new irons as material restrictions permit, but there aren't enough to go around.

If you already own a Coleman Iron—or any other Coleman appliance—keep it in tip-top working condition. Your Coleman Dealer can help make it work like new—with extra years of service for you. This will save you real money; and help conserve vital war materials, too. See your Coleman Dealer today!

THE COLEMAN LAMP AND STOVE CO., LTD., TORONTO, CANADA

No. 9637

Advertisement for the Coleman Lamp and Stove Company, 1942.   BUTTER-FAT, APRIL 1942

Advertisement for BC Electric Railway, 1926.   BUTTER-FAT, AUGUST 1926

No Milk Cheque

My father and mother immigrated to Canada in 1947 sponsored by one of my uncles who owned a dairy farm on Sumas Prairie at the time. Several of his brothers had moved to Canada in the late 1920s and they were all involved in small retail dairies up in Trail. Dad actually started farming on January 1, 1948 and started shipping to Melrose Dairy. By the middle of March, Dad hadn't received a milk cheque yet and we got a letter saying that the dairy had gone belly up so there would be no money for milk shipped up to that point. Then Dad shipped to Palm Dairy and later to the FVMPA.

Harry DeJong, 2000.

culture, Dodsley Barrow, got the Sumas Lake reclaimed and all the marginal land dyked. It made Chilliwack.

I remember one time when I was eight years old, my dad had to go to Victoria on a dyking business. He was a dyking commissioner. There were two factions, like there always is — one that wanted the dyke and another that didn't want the dyke. He took me out of school for the trip. We got on the boat at Sumas in freshet time. It was the Beaver, the best paddlewheeler. We got on at 8:00 a.m. in Sumas Landing and we made awfully fast time. We came to Mission, and the train was partly on the bridge, but they had the bridge open for the boat to go through. We sailed through there and we were down in Westminster at 12:00. It was an awfully fast trip. We got on the Interurban — the Central Park Line. I think you could ride from Westminster to Vancouver for twenty-five cents. When we got off, we had some time in Vancouver. Two other men were with us and they took me into a restaurant to have something to eat. And my dad wouldn't eat because there was quite a wind blowing and knew it would be pretty rough sailing. I went in and had a meal, ham and eggs or something. We started out to Victoria, and just got out into the Gulf and I had to make a hightailed trip out to feed the seagulls! We arrived at Victoria about 10:00 at night. The next day we went to the Parliament Buildings and I shook hands with Premier McBride under the dome of the legislature. He was Premier for about 16 years. He said that probably one day I would be Premier, but I didn't want that kind of job! We came home by boat, too. It took quite a bit longer.

In 1910, 1911, and 1912, BC Electric put the

Harry Irwin

train and electricity into Chilliwack. And that was a godsend. You could run your separator with a motor. We got electricity right away. At that time, too, the farmers started to ship milk to different dairies at Vancouver, after the BC Electric put on a milk train. It used to pick the milk up at different roads. The CNR went through 1911 or 1912 and they put on a milk train that started at Hope. It used to pick milk up at different roads and it was quite a fast train and passengers could ride on it too. It made Vancouver next door. But the greatest thing was when they reclaimed the lake and all that marginal land, eliminated the mosquitos, and put all that land into production. It was a great move to the dairy industry.

I had a mixed herd at the beginning. I left the farm and went to the coast and was in the trucking business. I had an accident — hit a train — and was laid up for over two years. I always kept my chin up because I had a beautiful wife and two wonderful children. I had a lot to live for. It doesn't matter how bad you are

Maple Springs Guernseys, R.H. Irwin and Son Farms, Dewdney, 1950s

Feeding People

When you feed people, you
feel good.

Mrs. Darshen Bains, 2000.

*down, you always want to be positive. Never be nega-
tive. So, my wife and I decided to go back to the farm.
I was hurt on the nineteenth of June 1930. I went
back to the farm in 1933. Money was awfully scarce.
Stewart Dixon was the cow testing supervisor, and I
knew Stewart quite well. Whenever there was any good
cattle, from herds that were on the Cow Testing Association, he would let me know. I started out with four
Guernsey cows, because I wasn't strong. Then I bought
young cattle from some of these herds, you see, and I
put them on test and right off the bat, I had a herd of
an average of 500 pounds. Stewart told me about*
Willow's Westerner, *just a calf. He said he should be a
good herd sire. So, I bought him and my son Robert
put him in the Calf Club. The third year after I got
him, I had seven heifers come into production. And Joe
Wingrove was the supervisor of cow testings who had
run the tests. I was working in one of the fields and he
come running across the fields, all excited. I wondered
what was wrong. And he said, "Harry! All the heifers
tested over 5 percent!"* Star, *one of them, went on and
as a two-year-old, produced 632 pounds of butterfat.*

*In a couple of years, we had to get another herd sire,
so, we went to Chuckanut Farms. Mr. Bartlee was
known as the "Daddy of the Guernseys" in Washington.
He was a great judge of cattle and made a lot of world
records. Mother and I and Joe Wingrove, the cow tester,
made a trip to Chuckanut Farms.* Miss Alberta *was
running in the loafing barn. She had a wonderful
udder, just like an Ayshire. Her son was in the calf
barn and so we looked him over. We went back into the
house and went back through the pedigree for eight*

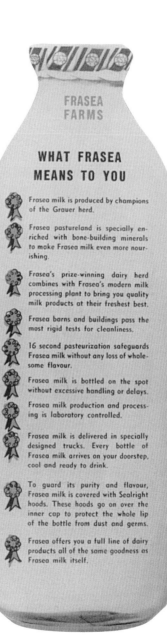

**Frasea Farms
brochure, circa 1950.**

234

Maple Springs Farm, Dewdney, after the Hatzic break, 1948.

Carrying the Can

People used to haul their milk out to a stand on the road or to the railway stands. When my father first started, he didn't have many cows, but he carried the milk can on a Chinese pole across his shoulders to the train. There were some stumps cut off along the way, and once in a while he could put the can down on a stump and rest and then keep on going. Eventually we got a road built and we hauled by horse.

Gordon Park, 2000.

generations. And Joe Wingrove *said, "Get him Harry! Get him!" So, I said to Mr. Bartlee, I said, "Well, I want to take* Chuckanut Elmer." *At which he said, "I don't think you'll make any mistake. And if he doesn't pan out, you come right back to me." He proofed out to become one of the greatest sires in Canada. I had seven of his daughters, they were on all on the honour roll for years.* Flossy *made four records or five records of over 800 pounds. And* Charity, *who was the highest producing two-year-old in Canada for years, made 750 pounds of butterfat as a two-year-old.* Irene *made over four tons of butterfat. I think she had five records of over 800 pounds. They were wonderful cattle in the parlour. As soon as you washed and prepared their udders, the milk was there and you had to get the machine out right away. Well, they are nice memories.*

When I was building the herd, I never put out a good grade animal until I had a purebred as good or better. I had old Blossom, *for example. She was a great cow and she freshened every year and she made 565 pounds practically every year. She didn't quit until she was 17 years old. And she never gave a drop of off milk.*

I had good hired help. I had a boy, Jack Seamus, *who came from Ireland. He was the best stockman I ever had in my life. He had a way with cattle. If I wanted anything to get ready for a sale, he would get that animal in just perfect condition. He would say, "Give me your ring, lad!" and the bulls would come. I used to tether the poles out in the grass and this one day my wife came running out to where I was, she said* Dougall MacDonald *has just phoned and said that* Black Magic *is over in the back fields, the 40-acre*

fields. So, I forget what I was doing, got the staff and got the old border collie, at my heel all the time, and went across the fields, and Magic *was pawing the ground.* Dougall MacDonald *said, "Open the gate and let him into my barn area." I said, "No." I had the old dog behind me, and I kept walking towards* Magic *and kept his eye. I said, "Give me your ring. Give me your ring." I kept saying that. He quit his pawing, put his head up, put the staff through the ring, and started across the field.* Dougall MacDonald *said he had never seen that before. But it was Seamus's work. I went into the calf barn one day, and there were about a dozen calves all running in the driveway, I said, "What are you doing Seamus?" And Seamus said, "Oh, I'm just giving the calves a bit of exercise."*

I shipped to Guernsey Dairies, *where I got a premium on the milk, until they sold out to* Jersey Farms. *We moved to Dewdney in 1947. I studied the map and that was all Monroe Loam, all that district. It was the best farm land you could get. I bought the place and moved the cattle there. Then the 1948 flood came. The pumps blew up at the dyke and our place was right in the path of it. It took the house and all the buildings right out. I saved the cattle by a minute and a half. The bridge across the slough, I had the trailer loaded with calves, and I almost lost the whole works. I had the trailer loaded with calves, milking machines, and all my ROP papers all in the car. There was a real torrent tearing under this bridge, and the Public Works were there and they said, "Hit it, Harry!" I was taking it quite slow. I drove that car up to higher ground, walked back. They said, "Where are the cattle?" I said they were just coming*

Dairy farm near Golden, 1955.

The Milk Tester

The milk tester came once each month. He sampled and weighed each night milking, stayed at the farm overnight, then weighed and sampled each morning milking. The two-milking sample from each cow was then tested for fat and the weights of the milk and fat production for the month were calculated. The milk tester recorded all results in the test book before leaving the farm.

Bob Irwin, 2000.

around the corner there. There were a hundred and ten head being herded because I didn't have time to get them into trucks. Old Nancy, one of the best cows, always the first one in the parlour, she was the lead cow. She was the head of the whole works. I went out onto the bridge and said, "Come on, Nance! Come on!" She came on, the rest followed. A minute and a half after they got across, the bridge went out. We took them on for another four or five miles and there was a little settlement in an opening and had them rest there. They were all loose, running in the bushes. Luckily, I had bought a couple hundred feet of half-inch rope and this was a French settlement and they were awfully kind to us. We milked the cows and fed the calves. We gave milk to the citizens because they gave us sandwiches.

Three landing barges came up from the Coast, up the Fraser and moved the cattle over to higher ground, about six miles. A hundred and ten head and they were loaded into trucks and they were taken into the Mission Fair Grounds where there was about two hundred head. My wife was at the coast. My children, Marie and Robert, were with me. Murray Davie was working on the dykes in Delta — they had the advantage that they could work on the dyke while the tide was out. Murray phoned my wife and said, "Tell Harry to ship all these young stock and all milking cows down to me and we'll look after them." So I got a big cow truck and loaded up the whole works and then went to Del-Eden. Murray and Ken looked after them. The barn wasn't finished and some of them were going to calve. Murray kept them in the barn and feed them grain. They were good people. He wouldn't take a cent for it. ■

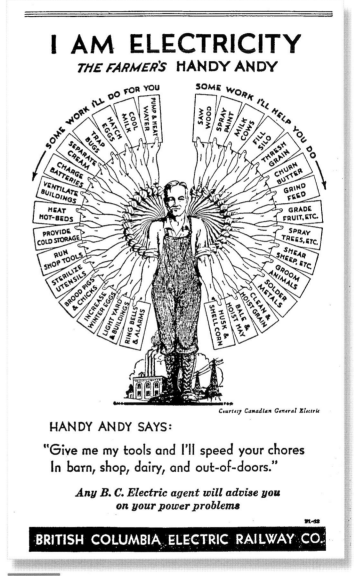

Advertisement extolling the virtues of electricity, 1932.
BUTTER-FAT, FEBRUARY 1932

238

One of Van de Hoop's prize Ayrshires, Pemberton, 1921. Note snowbank and bucket in foreground.

CHAPTER VIII

The Farm

Barn raising, Davidson farm, Fort Langley, 1905.
BCARS C09092

A Birthday Present

During the 1930s, both my grandfather and my dad had their own respective farms in Richmond. They were small farms in those days — my grandfather had a 20 acre farm with about 8 or 10 cows. My father had a 40 acre farm with about 12 or 14 cows. These had to be milked by hand and wives and sons were expected to help with the milking. For my eleventh birthday, I received a two-gallon milk pail as a present.

David Blair, 2000.

1937 White 3-Ton truck, Kosiancic Farm, Crescent Valley.
RAY KOSIANCIC COLLECTION

1919 Forage Cutter, Kosiancic Farm, Crescent Valley.
RAY KOSIANCIC COLLECTION

Loose Hay, Kosiancic Farm, Crescent Valley.
RAY KOSIANCIC COLLECTION

Gifford Thompson making silage, Kelowna, 1957.
LLOYD DUGGAN COLLECTION

A great many changes on the farm have happened to accommodate advances in technology and related changes in farm management philosophy. Early barns were multi-use buildings, often built of logs or readily available material. Sometimes logs were put aside as the land was cleared, then hauled away to the local sawmill and cut into lumber before being returned to the home place for building to begin. Glimpses of the old barns in the pages that follow remind us that building and threshing times were also community times — times when labour, machinery, and knowledge were shared amongst neighbours to get the job done. ■

Potato planting, Basil Gardom farm, Dewdney, 1930s.

Dryness

Dampness is a curse in any stable. It makes the cows delicate, breeds disease, is most unpleasant for the farmer himself, and ruins implements or harnesses. The correct number of cubic feet of air space for each animal, a proper ventilating system, and walls built in such a way as to prevent condensation will keep the stable dry. This is one of the most important essentials.

Beatty Barn Book, 1937.

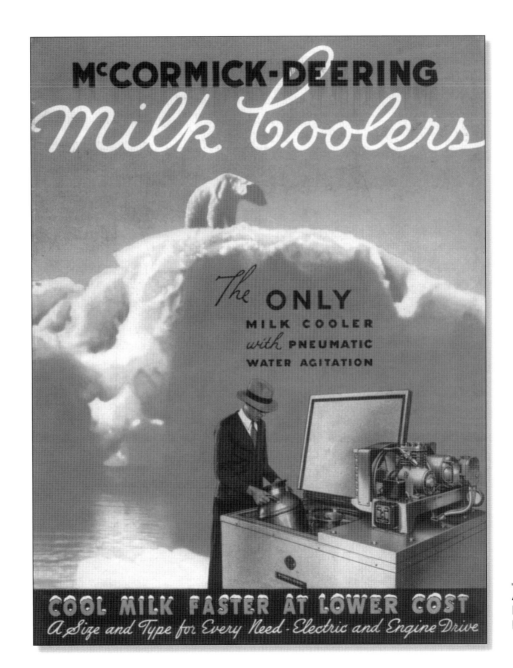

Advertisement for McCormick-Deering Milk Cooler, circa 1950.
ACTON KILBY COLLECTION, KILBY FARM MUSEUM

244

Work Horses, Belmont Farm, Langley, home of the Berry Family, circa 1910.

Silage cutter, electric motor, and blower for filling silo, J.W. Berry's Belmont Farm, Langley, 1917. The silage cutter and blower were owned by Mr. Berry and his neighbours on shares according to their acreage under cultivation. Similar installations in the area were at the farms of the Hagarty Brothers, B.A. Harrison, and A.R. Webster; at the farm of Charles E. Hope in Fort Langley; at the farm of Charles Reid in Chilliwack; and at the Shannon Brothers farm in Cloverdale.

Milking Machines

Milking machines, with vacuum pumps powered by gasoline engines or electric motors, were in use on some larger farms in the 1920s. The rural electrification of farms in the 1930s made milking machines more available.

Bob Irwin, 2000.

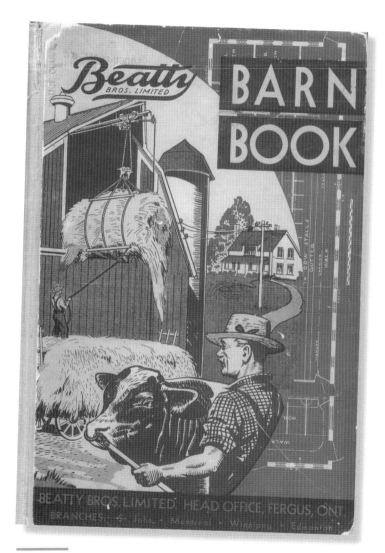

Beatty Brothers Ltd. supply catalogue, 1937.
WALTER GOERZEN COLLECTION.

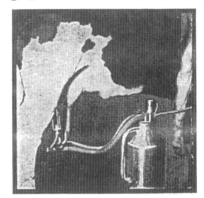

60 FARMERS

CAN'T BE WRONG

Last year over 60 new milking machines were installed on Fraser Valley farms. Modern electric milkers operating quietly, efficiently, cut production costs way down. Cows give more milk if they are milked at exactly the same speed and pressure each operation, and milking time is cut at least in half.

Sixty Valley farmers declare that no other piece of electrical equipment can save more time and money than a milking machine.

Your B.C. Electric representative will be pleased to answer any questions you desire concerning electric milking.

BRITISH COLUMBIA ELECTRIC RAILWAY CO. LTD.

BC Electric advertisement, circa 1930s.
BUTTER-FAT

Threshing at Savage farm, Richmond, 1915.
JAKE SAVAGE COLLECTION

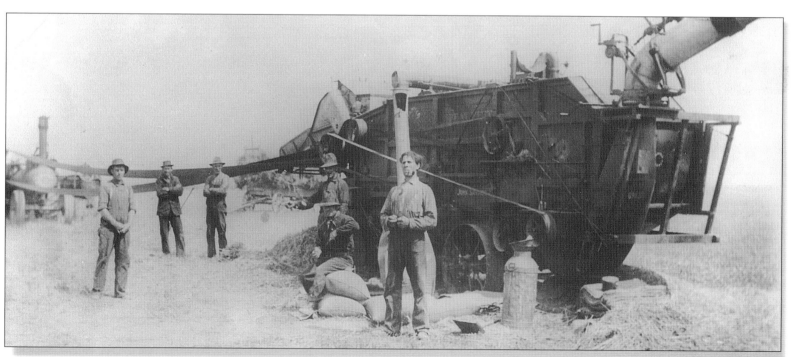

Threshing at Savage farm, Richmond, 1915.
JAKE SAVAGE COLLECTION

Appearance

It is well worth while to give some thought to the outside appearance of the barn. It is not a costly matter to make the barn look attractive. Put dormer windows in the roof, and add a cornice and cupola. Paint the barn. This is necessary in any case to preserve the building, and adds to the appearance as well. Remember that a handsome appearance has a cash value.

Beatty Barn Book, 1937.

Plowing, Lulu Island, 1913.

Baling Hay at Michell farm, Saanich, 1934.
BCARS G-02946

Barn at the farm of the Consolidated Mining and Smelting Company (COMINCO), Tadanac, 1930.
BCARS B-05032

New Ways

Ernie Winterhalder says he positively would not be in the dairy and hog business if he did not have his milking machine. He has used one now for three years. He got it from the Enderby Creamery.

Once in three years he couldn't start the engine because of a defective spark plug. He had to milk by hand and turn the separator by hand, too. He says he didn't finish his chores until ten o'clock in the morning.

"It sure got me," he said with a big smile.

The Cream Collector,
March 1945.

"BLUE-RIBBON"
The Popular Milking Machine

Authoritative estimates place the total number of electric milking machines sold during the past year in the Fraser Valley as 60. During the past year FORTY-ONE of these were BLUE RIBBON MILKERS, or more than two-thirds. This reputation was not earned in a day, or a year,—but over a period of thirteen years during which these practical and economical milkers have been serving Fraser Valley dairymen.

FORTY-ONE
FARMERS
CAN'T
BE
WRONG

COSTS—The cost is low because there are no expensive and cumbersome pipelines to install and maintain. BLUE RIBBON MILKERS sell for ONE price and ONE PRICE ONLY.

For the Single Unit $140.00
Double Unit $207.00

Write to
Fraser Valley Farm Specialists
BOX 500, CHILLIWACK, or PHONE CHILLIWACK 3441
for free descriptive literature

Advertisement for Fraser Valley Farm Specialists, 1940. BUTTER-FAT, MAY 1940

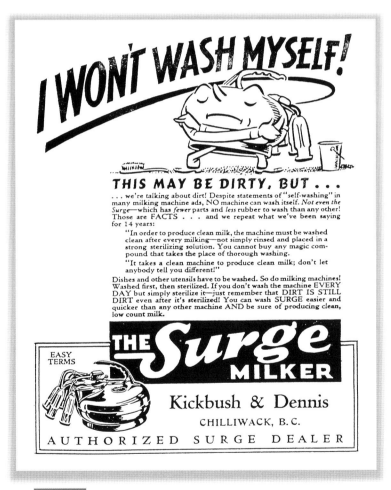

Advertisement for Surge milker, 1940. BUTTER-FAT, SEPTEMBER 1940

Elevated can mover, J.M. Steves Dairy, Richmond, circa 1930.
RICHMOND MUSEUM AND ARCHIVES, 1978 73

Barn with manure management system, Basil Gardom farm, Dewdney, 1930s.
GARDE GARDOM COLLECTION

Hurrah!

It's coming. Rural electrification is on the way. The provincial government has indicated that legislation will be brought before the 1945 legislature which will open the road to complete rural electrification.

Power for 1415 Okanagan farms now without it will mean water pressure at the tap, modern bathrooms, modern kitchens with electrical labour-saving devices, refrigerators, etc.

The greatest step forward in improving rural living conditions is now to be taken.

Not much longer will farmers have to wash in a hand basin, bathe in the wash tub, run out to the toilet. The bad days are over when the farmer's wife must lug in water, iron over a hot stove, or do without modern electrical conveniences.

The Cream Collector,
February 1945.

This Booklet awaits your address

We want you to have this interesting booklet whether you are thinking about saving milking time or not. Simply send postal card and you get it by return mail without incurring the slightest obligation.

Tried, Tested and Approved by Canadian Farmers.

LISTER MILKING MACHINE

Certainly you admit it is worth any man's while to learn how to increase his herd and have more dairy products to sell without extra labor. This booklet tells you how: Farmers pay for the Lister Milker out of extra profits. Simple. Anyone can operate. The Lister reputation is behind it. Also ask about the famous Lister Engine. When writing ask for Booklet A8.

R. A. LISTER & CO., (Canada) Limited
58-60 Stewart St., Toronto

Advertisement for the Lister Milking Machine, 1920.
FARMERS' MAGAZINE, 16 OCTOBER 1920

Treadmill used to power milking machine and grinder, Belmont Farm, Langley, 1902. Harry Berry is sitting; J.W. Berry is standing.

CHUB BERRY COLLECTION

Save Valuable Haying Time

A half day gained in the race with the weather may save you many tons of valuable hay. And it doesn't take much hay to pay for an outfit of Louden Hay Loading Tools. Two men and a boy with these tools can do the work of ten.

Louden Hay Tools are simple, no complicated mechanism to cause costly delay. Write us for complete illustrated catalogue.

A. I. JOHNSON & CO., Ltd.
844 Cambie St.   Vancouver, B. C.

Advertisement for Louden Hay Loading Tools, 1926.

BUTTER-FAT, MAY 1926

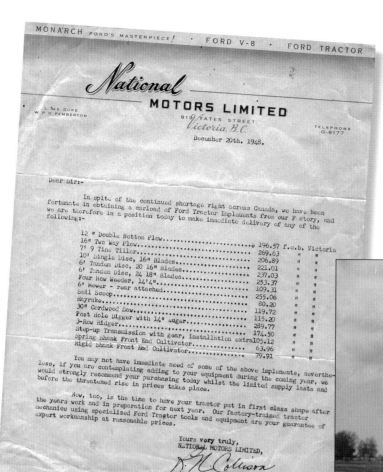

Quote for farm equipment, National Motors.
JOHN PENDRAY COLLECTION

First rubber-tired seed drill in the Fraser Valley, Savage farm, 1946.
JAKE SAVAGE COLLECTION

Belts and Pulleys

The first milking machine we had was when I was about five or six years old. Ours was run with a big gas engine. Oh! There were belts and pulleys everywhere. It just dragged the milk out. It didn't work very well. We finally got a Surge, but every once in a while, the cow would kick it off and kick it all over the barn.

Glenn Toop, 2000.

Savage barn, Richmond, circa 1920.

JAKE SAVAGE COLLECTION

Casanave farm, Courtney. Jim Casanave, Sr. standing and John Casanave on tractor, 1947.

JIM CASANAVE COLLECTION

Holsteins lined up for milking at Raper Farm, Colquitz, 1947.
BCARS I-21141

Brackenhurst Dairy Farm, Sidney, 1946.
BCARS I-26693

255

CHAPTER IX

The Herd

Cows owned by Fred Corbett in Sumas River, Abbotsford, early 1940s.
WALTER GOERZEN COLLECTION

Colony Farm

The first artificial insemination of dairy cattle in British Columbia took place at Colony Farm in the early 1940s. Colony Farm, a provincially administered program, was at that time located within the Department of the Provincial Secretary and was closely associated with the operation of the adjacent provincial mental hospital. The main technician involved was the herdsman, Spence Stroyan, who left the farm in the later 1940s to become employed by the newly formed British Columbia Artifical Insemination Centre.

Bruce Richardson, superintendent of government farms, including Colony Farm, from Fifty Years of Artifical Insemination in Canada, 1934–1984, *by Ron Snyder.*

Many changes have taken place in the dairy animal and in the dairy herd since the turn of the century. These changes have been made possible by meticulous attention to detail. Careful record keeping as an industry norm began with the first cow testing associations in the province, started in Chilliwack and Langley in 1913. These records enabled farmers to make informed decisions about selection. Since the Second World War, advances in science have made possible other changes. Diseases like tuberculosis and brucellosis, once endemic, have now been virtually eradicated. Nutrition has improved. Artificial insemination has become an industry norm. Other changes, especially those related to cow conformation, have been in response to changes in technology. The use of the milking machine today, in place of the hand milking of the past, has required changes in udder conformation: long teats suitable for easy hand milking have been replaced by shorter, neater teats to better suit modern milking machines. Changes in barn configuration and in animal housing arrangements have meant that different characteristics for feet and legs have come to be favoured to accommodate cows whose lives are no longer lived on pasture, but rather on concrete with minimal shavings. By far the most dramatic change in the dairy cow over the last one hundred years, however, has been in her sheer ability to produce milk.

Probably the single biggest management tool of the twentieth century for dairy farmers has been the development of artificial insemination (AI). AI began in Canada in the 1940s, and by the 1960s and 1970s had become a respected, reliable, and tested technology. It made possible a measure of safety on the farm, improved disease control, and gave farmers access to a wide range of bulls.

The roots of artificial insemination in Canada go back to February of 1936 when the first calf conceived through AI was born on the Central Experimental Farm in Ottawa.[1] In British Columbia, some early insemination was performed at Colony Farm, Frasea Farm, the Dominion Experimental Farm in Agassiz, and Stanhope Farm, owned by Ralph Rendle and family in Saanich.

In early 1944, interest was high enough among milk producers that farmers on both ends of the Fraser Valley resolved to form societies dedicated to AI. In May, the Chilliwack Artificial Insemination Club was formed and its first directors elected, with Harold German as president. Other members were Mike Goriak, A.J. Bailey, George Challenger, Jack Mace, A.S. Barker, and Ian Hepburn. H.C. Clarke served as secretary from 1944 to 1959. According to Roy Snyder, in *Fifty Years of Artificial Insemination in Canada*, this club "started its operations on January 19, 1945, with seven bulls leased from the Dominion Government. There were about 200 members, each of whom paid a ten dollar membership fee. With this money, the club purchased material for pens and a shed for a lab. Jack Andrews provided the space on his farm and members erected the buildings. The first manager, Tully McLean, was the only employee in the beginning, but Jack Andrews assisted in handling bulls."[2]

The Lower Fraser Valley Artificial Insemination Association was formed in September of 1944 with Tom Berry as president and F.J. Kellaway as secretary-

CANADA

CANADIAN RECORD OF PERFORMANCE FOR

PURE BRED DAIRY CATTLE

RECORD
OF

HOLSTEIN-FRIESIAN COW

COLONY LASS PAULINE HEILO _____ REG. Nº __233129__ R. OF P. Nº __16734__

SIRE __Hazelwood Heilo Sir Bessie__ _____ REG. Nº __66155__

DAM __Colony Lass P. Koba__ _____ REG. Nº __137922__

OWNED BY __ST. MARY'S MISSION, MISSION CITY, B. C.__

BRED BY __Colony Farm, Essondale, B. C.__

CLASS __Mature__ _____ AGE AT COMMENCEMENT OF TEST __5__ YEARS _____ DAYS

DATE TEST COMMENCED _____ March 27, 1936 .

DATE OF CALVING _____ March 27, 1936

DATE OF CALVING FOLLOWING TEST _____

MONTHS	LBS. MILK	LBS. FAT	MONTHS	LBS. MILK	LBS. FAT
JAN.	1036.0	30.14	JULY	1870.0	59.84
FEB.	957.0	27.75	AUG.	1512.0	48.38
MAR.	752.0 266.5	21.80 8.26	SEPT.	1388.5	44.43
APL.	2110.0	67.48	OCT.	1413.0	42.39
MAY	2083.0	72.91	NOV.	1265.0	37.95
JUNE	1955.5	66.67	DEC.	1141.0	34.23

PRODUCTION REQUIRED FOR REGISTRATION __12,000__ LBS. MILK __408__ LBS. FAT

TOTAL PRODUCTION __17,750__ LBS. MILK __562__ LBS. FAT

AVERAGE PER CENT OF FAT __3.17__ Nº DAYS IN MILK __365__

Geo B Rothwell
LIVE STOCK COMMISSIONER

C. S. Wood
CHIEF INSPECTOR

DATE __April 9th, 1937__

Milked three times daily for 219 days

ROP record for *Colony Lass Pauline Heilo*, owned by St. Mary's Mission, 1937.
BC HOLSTEIN MUSEUM COLLECTION

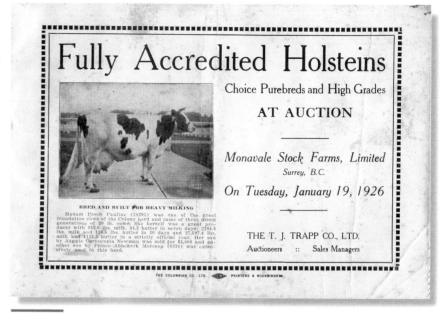

Monavale Stock Farm auction catalogue, 1926. BC HOLSTEIN MUSEUM COLLECTION

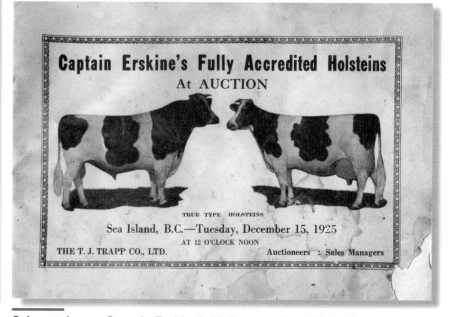

Sale catalogue, Captain Erskine's "fully accredited Holsteins," 1925.
BC HOLSTEIN MUSEUM COLLECTION

Moving Away from the Coloured Breeds

At one time, almost all of the cows in this area were coloured breeds. Jersey, Guernsey, and a few Ayrshire herds. There was the odd Holstein herd but not too many. Then the Holsteins came back to life when the Dutch immigrants came in. They really went to work on the Holsteins and showed us the way. Now there are very few of the coloured breeds. People realized that with Holsteins, properly managed, and with the proper genetics, you could get more butterfat in a year than you could with any of the coloured breeds. The Dutch did a marvellous job in improving the whole industry. They worked hard. I just can't say enough good things about what they did for the dairy industry of British Columbia.

Allan Toop, 2000.

treasurer. Harry Berry became president the next year and remained in that position for the next 15 years. The Association began its operations in Surrey in the fall of 1945 in a milk house renovated to accommodate a lab and office and with bulls on loan: two Jersey bulls, two Guernseys, two Holsteins, and one Ayrshire. In 1948, land was purchased in Milner, and a new barn, lab, and office were built and officially opened the following year. Beef sires were added to the centre in 1950. Both the federal and provincial governments played significant roles in these early years, as did the first manager of the association, Spence Stroyan. Dr. J.C Bankier, Livestock pathologist and BC livestock supervisor, was loaned by the provincial government to the association for a period of a few years to "devise ways of improving the conception rate and to solve financial problems."[3]

Other societies developed in the province at this time. Papers of the Dairy Branch indicate that the North Okanagan Artificial Breeding Club began operations in 1944, with an office in Enderby covering Armstrong, Salmon Arm, Enderby district, Grindrod, Mara, Deep Creek, Grandview Bench, Ashton Creek, and Trinity Valley. A letter from F.C. Wasson of the Department of Agriculture in Kelowna, to Dairy Commissioner Henry Rive, dated November 14, 1944, explains that there is "no danger of the movement falling through" because the club is "fixing up the old Jones place about one mile north of Enderby" and that it hopes to get three or four bulls to start with, including a Jersey, and Ayrshire, and a Holstein.

On Vancouver Island, the Cowichan Agricultural Association started an artificial insemination organiza-

tion in August of 1945. In the spring of 1946, "the first calf in the Cowichan district was born through AI. Later that year three small calves, the result of AI, were displayed in pens at the Cobble Hill Fair and at the Red and White Show. Great interest was shown by the public. Visitors were surprised that the calves looked no different from the other calves. They were referred to as test tube calves."[4]

Not all farmers accepted this new technology. In 1948, the *Langley Advance* reprinted two angry letters from English newspapers. The first suggested the practice "will mean untold torture to our animals — apart from mocking God and nature."[5] The second suggested that "the whole thing is as filthy a scheme as ever originated … and anyone engaged in practising the vile cult should be debarred from mixing with decent human beings."[6] The majority of farmers, however, were interested in the possibilities of this new technology and curious about its outcome.

The early years of AI were years of developing insemination technology as well as of educating farmers about the specifics of bovine reproduction and about the role that good nutrition and disease prevention played in conception rates. The BCAI Centre started the Young Sire Proving Program in 1954, the first program of its kind to be developed in Canada. In accordance with this program "most bulls entering the program are the result of planned matings … an arrangement between a breeder and the AI unit whereby the breeder agrees to mate one of his best cows for production and type to a selected proven sire. He grants the AI unit the option of purchasing the resulting calf, if it is a bull."[7] This for-

CANADIAN NATIONAL RECORD

Holstein-Friesian Association of Canada.

Incorporated under the Act respecting Live Stock Record Associations at the Department of Agriculture, Ottawa, Canada.

Certified Copy of Pedigree Recorded in the Canadian Holstein-Friesian Herd Book.

No. 63447.   Vol. 23.

Female   Bonnie Moonface of Milner

Owned by   and

Bred by   William Hunter, Milner, B. C.

Calved   April 18, 1917.

Color, black and white.   Diagrams on reverse of this Certificate.

Sire   Calamity Abbekerk Boy   No. 20289.   H. B.

Dam   Lady Bess of Milner   No. 18643.   H. B.

January 21, 1919.

W. G. Clemons Secretary and Registrar

This is to Certify that the above pedigree is on record in Volume 23.

This slip, when presented with and attached to the certificate, will be authority for the agent of the railway company to waybill at the reduced rates agreed to with the Dominion Department of Agriculture. This slip must be detached from the certificate by the agent and forwarded with the waybill.

FEMALE   Holstein-Friesian Railway Shipping Voucher No. 3.   Age..........

Name of Animal ....Certificate No..........

Signature of Shipper....................P.O.

This slip, when presented with and attached to the certificate, will be authority for the agent of the railway company to waybill at the reduced rates agreed to with the Dominion Department of Agriculture. This slip must be detached from the certificate by the agent and forwarded with the waybill.

FEMALE   Holstein-Friesian Railway Shipping Voucher No. 2.   Age..........

Name of Animal....................................Certificate No..........

Signature of Shipper....................P.O.

This slip, when presented with and attached to the certificate, will be authority for the agent of the railway company to waybill at the reduced rates agreed to with the Dominion Department of Agriculture. This slip must be detached from the certificate by the agent and forwarded with the way bill.

FEMALE   Holstein-Friesian Railway Shipping Voucher No 1.   Age..........

Name of Animal....................................Certificate No..........

Signature of Shipper....................P.O.

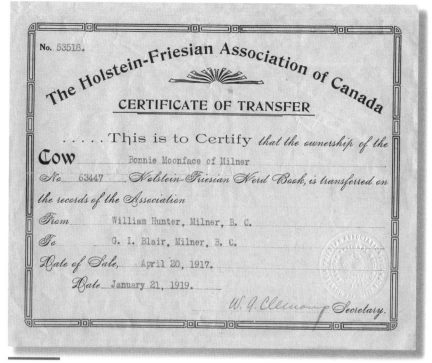

The Holstein-Friesian Association of Canada

No. 53518.

CERTIFICATE OF TRANSFER

.....This is to Certify that the ownership of the

Cow   Bonnie Moonface of Milner

No. 63447   Holstein-Friesian Herd Book, is transferred on the records of the Association

From   William Hunter, Milner, B. C.

To   G. I. Blair, Milner, B. C.

Date of Sale,   April 20, 1917.

Date January 21, 1919.

W. G. Clemons Secretary.

Transfer certificate, *Bonnie Moonface of Milner*, 1917. This cow was a foundational animal in the Blair family herd. In 1923, she produced 22,000 pounds of milk and 1,000 pounds of fat.

BC HOLSTEIN MUSEUM COLLECTION

Bonnie Moonface of Milner, registration papers, 1917.

BC HOLSTEIN MUSEUMCOLLECTION

AI in Richmond

I was living in Hammond in 1947, and I was asked to come down and do some infertility work at Frasea Farms. Then Les Gilmore came to me in about 1948 and he said he was going to buy a bull with McKims, and they wanted to get *Carnation Skipper*. They paid a lot of money for him but they couldn't use him for breeding cows — they had to do it all artificially. So I put up a program where I collected the semen and artificially inseminated the cows at the three Gilmore farms and at the McKims. Along with doing artificial insemination, I also did a lot of infertility work too, so that really kept me busy. That's how I got started.

Dr. Joe Lomas. 2000.

malized bull testing program, pioneered by the BCAI Centre, allowed the industry for the first time to gather enough data on bulls to begin to assess their genetic worth in dairy herds.

Today, the territory of the BCAI Centre has expanded as has its use of new technologies, but its basic role of offering genetic services to livestock breeders across western Canada has remained the same. In 1997, its name was changed to Westgen to reflect its geographic role in the new Semex Alliance — a group of affiliated AI organizations across Canada.

Dairy Farm Development in the Creston Valley
Sig Peterson, 2000.

The Kootenays, in a general sense, were remote from the major dairy centres in the province. Transportation other than by rail meant travelling by road through Washington State to the coast. I think none of the producers had ever been on modern dairy farms in the Fraser Valley. There was no road to speak of from Christina Lake to Castlegar. It was an isolated situation with little outside farmer-to-farmer dialogue, which is essential to accelerate the rate of progress. The isolation also made it very difficult to find quality breeding stock locally for herd improvement. The only significant herds of quality were located in Trail and Kimberley and operated by the CMS mining company. These were purebred Ayrshire cattle, not often available for local purchase. I observed on my visits to dairy farms that the cattle were of mixed breeding. White faces or other beef breed markings were common; they were practical cows for cream production and for beef.

It was apparent that we would have to import stock from the coast or the Okanagan to improve the milking herds. AI services were unheard of and not likely to become available soon.

The Livestock Branch of the Department of Agriculture was available to assist us. I wrote a letter to Frank Clark, a staff officer in the Branch that described our situation at Creston in considerable detail and told him that we were not looking for purebred cattle; we only needed bred Ayrshire grade heifers from good producing herds. Frank was very knowledgeable about our needs and about sources of good stock. He replied that he would have a carload of the kind of heifers we needed at good prices within a few weeks time. His assistance led to the arrival of a carload at the siding in Creston. I accompanied the farmers to greet the new arrivals with a good deal of inner excitement, wondering if the heifers would be up to what the farmers had been led to believe by myself. They were indeed good looking stock and no difficulty was experienced in allocating them to each of the buyers. Also, at that time I learned of a purebred Ayrshire herd sale in the Okanagan from Clark. Two of our farmers drove to Armstrong and bought six of the best cows in the herd. When I think of the initial carload now, it says so much about the calibre of the dairymen who trusted me when I assured them that Frank Clark would pick out really good grade stock and have them shipped as soon as I forwarded their certified checks in payment. Those early imports, followed by others several months later, laid the early foundation of a much improved dairy herd in the Creston Valley. It doesn't seem like

British Columbia Department of Agriculture
(Live Stock Branch)

LIVING TON-OF-BUTTERFAT COWS

A list of 498 cows on test in provincial cow-testing associations as at June 30th, 1945, each of which has production records totalling a ton of butterfat or more.

When reporting pure-breds, barn names are given together with their registered numbers. The association of which a dairyman is a member is indicated by a designating letter: Chilliwack, C.; Delta, E.; Dewdney-Deroche, N.; Langley, L.; Matsqui, M.; Richmond, R.; Salmon Arm-North Okanagan, D.; Okanagan, K.; Sumas, A.; Surrey, S.; Vancouver Island (South), V.

JERSEYS

Name of Cow	Milk Lb.	Fat Lb.	No. of M.P.	Owner & Association Letter	
Violet	97,237	4,620	9	A. Arthur	C.
Mystery	95,285	4,583	10	A. S. Barker	C.
Daisy	91,479	4,460	12	C. R. Newby	C.
Darkie	93,678	4,423	10	G. A. Swan	V.
Bunny	86,163	4,254	7	J. C. Brannick	C.
Rose 3rd, 82744	83,224	4,234	9	F. J. Appel	N.
Rose 4th, 110709	91,223	4,154	9	F. J. Appel	N.
Natalie, 88369	88,371	4,047	9	R. Machell	M.
Lady, 95279	74,417	4,042	8	Easton & Goodwin	V.
Fritz	80,853	3,964	9	Wm. Bothwell	S.
Trixie	91,462	3,963	12	T. Pennington	N.
Marg.	78,942	3,927	9	H. T. Tweddle	C.
June	63,285	3,903	7	J. O. Grigg	C.
Polly	60,581	3,715	10	D. Jones	D.
Lily	76,039	3,661	9	J. T. Godfrey	V.
Collen	79,092	3,653	9	J. O. Grigg	C.
Tulip, 80862	76,694	3,644	8	R. Machell	M.
Sophias Belle, 74407	73,192	3,633	9	E. Anderson	M.
Matilda, 91393	68,328	3,632	9	R. Machell	M.
Sunrise	76,407	3,628	8	G. A. Philp	C.
Martha	74,245	3,606	7	W. W. Elliot	R.
Mabeline, 98946	66,867	3,598	9	R. Machell	M.
Sibb, 86835	65,838	3,520	8	Easton & Goodwin	V.
Sugar	71,983	3,501	7	Forslund Bros.	L.
Spot	68,093	3,397	8	D. Jones	D.
Pansy	60,105	3,387	8	Forslund Bros.	L.
May	68,859	3,357	7	A. Glintz	V.
Flo	60,684	3,342	9	W. C. Boss & Sons	D.
Curly	73,779	3,322	8	H. T. Tweddle	C.
Meek	72,302	3,302	9	Thomson Bros.	C.
Dolly 2nd	57,633	3,208	6	Easton & Goodwin	V.
Joyce	64,325	3,200	8	A. W. Pearse	C.
Cecelia	55,325	3,165	7	A. Glintz	V.
Ramona, 105751	53,152	3,152	6	Easton & Goodwin	V.
Buttercup 2nd	74,381	3,137	8	R. W. Mercer	V.
Dolly	67,810	3,121	8	T. Smith & Sons	M.
Molly	55,490	3,062	7	J. O. Grigg	C.
Pet, 94233	63,611	3,005	7	G. A. Swan	V.
Maggie	73,707	3,004	10	W. C. Boss & Sons	D.
Ida	64,548	2,975	7	A. Glintz	V.
Norah	60,983	2,962	8	K. B. McKechnie	D.
Betty	62,918	2,943	7	T. Pennington	N.
Duchess	57,883	2,941	7	A. W. Pearse	C.
Lummy Lou	54,676	2,938	6	G. A. Philp	C.
Blackie	51,707	2,914	6	A. Arthur	C.
Lily	61,394	2,911	8	R. Whipple	D.

Low Producers and High Producers

It boils down to the fact that it takes the same amount of work to look after a low producer as a high producer.

Dick Graham, 1993.

Butterfat records issued by BC Department of Agriculture, 1946.

much now, but I was happy then with the role I had played in helping to bring it about.

In addition to improving the dairy stock, there was a need to improve the forage quality for high producing cows. The Creston district had long produced quality alfalfa hay in quantities surplus to local needs. It was shipped out of the district but the returns were often marginal if hay was abundant in the Kootenays. The dairy herd was grazed on dryland pastures, mostly brome, timothy, and rye grass mixes. Part of my work was to put out test plots to demonstrate new varieties of grasses and legumes. Orchard grass was new to the district as was trefoil. Both produced well as pasture forage and lessened the risk of bloat which was prevalent when alfalfa was used for pasture. Little was known about silage of any kind and especially grass silage put up in pits. A few innovative farmers were persuaded to give it a try with good results; the practice was soon adopted by others able to purchase, share, or borrow the special type of machinery needed. Silage gave good results and contributed to increased milk production in the winter months. Creston reclamation lands provided an abundant source of cereal grains and peas from which supplementary rations were formulated. All of the fundamental needs of a strong dairy industry were in place if a stable fluid market could be developed.

At this juncture I should like to tell of an amusing personal experience when visiting a cream shipper in the Nakusp district. On entering the yard of his remote bush farm on a November day, my eyes fell on quite an astonishing scene. Several cows were chewing on something that looked like small green logs, two to three

inches in diameter and two to three feet in length. One would immediately think they were smoking "cow cigars" — stogies sticking out the sides of their mouths at all angles. As I stood there looking at the situation, the farmer emerged from his house to greet me. We had a congenial conversation about a number of things and I was soon ready to leave. My curiosity got the better of me and I could not go without revealing my ignorance of what the cows were doing. "Tell me," I said, "what are your cows chewing on?" He looked at me with a quizzical grin on his face and replied, "Don't tell me that you're one of those university fellers and you have no idea what their eating?" "No," I replied. "I've never seen this before in my life." "Kale," he said, "Kale is what they are eating." After a good chuckle at my expense he spoke further. "This is known as marrow stem kale. We used to grow it back in England and it is a fine staple for milking cows." I replied, "If you don't mind my saying so, they look more like they're smoking cigars; the only kale I know about grows about eighteen inches high and has broad leaves." I had learned something new that day so long ago, and I have never forgotten the entertaining sight of the smoking cows.

Dairy cows in the district seemed uncommonly subject to milk fever. Most dairy farmers could expect to have one or two of their best cows come down with it after freshening. Visiting experts were unable to explain why it was so prevalent, particularly, it seemed, in the Lister community. Whether it had something to do with the high percentage of alfalfa in their annual diets; or some element missing in the soil was a moot subject for discussion. The answer was never found in my time in the

Breeding and Feeding

When I farmed in Richmond we talked in terms of butterfat. I believe my dad was third in the Association and I think at that time he had a herd average of about 406 pounds of butterfat per cow a year. That would be back in 1941 – 1942. Before I sold the dairy herd in 1962, we had somewhere around 560 or 570 pounds of butterfat per cow on average. Today an average would be over 600 pounds of butterfat. This has been accomplished through breeding and feeding. The BC Artificial Insemination Centre at Langley has done an awful lot.

David Blair, 1994.

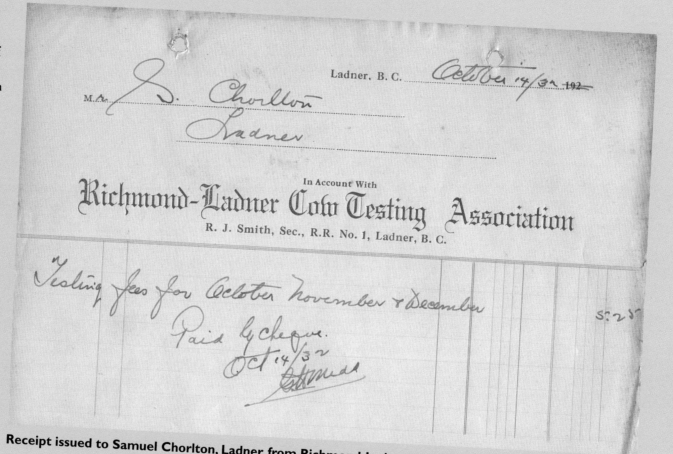

Receipt issued to Samuel Chorlton, Ladner, from Richmond-Ladner Cow Testing Association, 1932.

You've probably run into some of the old DHI records that are available. Back then, some of the cows hardly gave enough milk for a cup of tea. Our cows made 60 pounds of milk a day when I milked by hand when I was young. That was a top-notch cow. Now some do well over a 100 a day.

Glenn Toop, 2000.

district. The only treatment at the time was to inflate the cow's udder with a bicycle pump and hope you didn't introduce mastitis at the same time. Early in 1950, I had invited Dr. John Bankier, head of our veterinary laboratory, to come to Creston and speak on the subject of milk fever and mastitis. On the day following his talk, we visited several of the farms that had recently had the problem. By mid-afternoon Bankier had a full grasp of the situation and said to me, "Mr. Peterson, the fever is not difficult to deal with. It's a veterinarian's job, but I'm sure you could do just as well if you wanted to do it. Are you interested in learning how?" The next morning we went to the local drug store, bought a couple of needles, a yard of quarter-inch rubber hose and a bottle of calcium gluconate. We next called on a "fever farm" and the farmer was most willing to let me undergo a crash course on administering the solution to one of his cows through its jugular vein. Scarcely a week had passed when the phone rang in my office. "Sig, I hear you know how to inject a cow with that stuff that fellow Bankier was talking about at our meeting. My best cow is down with milk fever and if she doesn't get help in an hour or so I'm certain she's a gonner. Will you come out and see her?" I made haste for the drug store, bought a bottle of calcium gluconate and drove to the farm. There I found a beautiful Holstein cow lying flat out on her side. He had inflated her udder drum tight, but to no avail. Before taking a step nearer to the cow, I said to him, "Shake my hand — this is the first time I have done this and if this doesn't work, you're not going to hold me responsible for killing your cow." He quickly replied, "Go ahead and inject her, because if you can't help her she will die any-

way." With his assurance, I got to it and administered the solution. To our mutual astonishment, in about a half an hour she raised her head and began to look around. In another hour or more, she struggled to her feet and stood shakily on her own. My friend stood quietly with glistening eyes full of appreciation for my having saved his prize cow. Word quickly spread to other dairymen and you can easily guess what followed in the weeks and months ahead. I found myself administering the solution whenever needed and was glad to do so, although I always worried a bit about the outcome should one die. There was no charge for my services, of course, but I will admit to being given a bottle of cream or a freshly killed rooster as a token of thanks. A veterinarian was urgently needed in the district, but the livestock population was too small to support a service. I did my best to help out with ordinary livestock health problems until I left the district in 1953 on transfer to Courtenay.

Almost a half a century has since passed with enormous changes having come about in the Creston dairy industry. Dairy herds are now on a par with the best in the province, veterinarian services are readily available, the markets are well established, all of which brings joy to my heart. Great credit is due to those who persevered and brought the industry to its modern state of development.

The Beginnings of AI in British Columbia
Dr. A. Kidd, 2000.

When I started with the Department of Agriculture in 1947, the average cow on test was giving about 7,000

During the spring months of 1950, 1951, and 1952, the BC Department of Agriculture sponsored a series of educational programs, called "Planned Farming," held at farm centres in the Fraser Valley. Left (near microscope), Dr. F Morres, veterinary inspector; centre, Mr. F.C. Clark, livestock inspector; right, Dr. A. Kidd, veterinary inspector, Cloverdale, 1950.
DR. A. KIDD COLLECTION

Committee that drew up regulations for licensed dairy farms following the Clyne Commission. Regulations covered construction and sanitation of farm buildings, as well as animal health requirements. The committee comprised dairy farmers, public health officials, and agriculture officials. Committee executive, left to right: Vice Chairman John Kirkness, dairy farmer, Chilliwack; Secretary Dr. A. Kidd, Department of Agriculture, Victoria; Chairman Dr. D. MacKenzie, assistant dean of agriculture, University of British Columbia, 1957.
DR. A. KIDD COLLECTION

Dr. A. Kidd (left) and Graham Lambrick (right) vaccinating calf for brucellosis at the Lambrick farm, Saanich, early 1950s.
DR. A. KIDD COLLECTION

Progress

If ever anyone asked me what I thought was the most important contribution of the 20th century to human progress ... I should say without hesitation, the bull farm. Here you have your cow at one end of the phone, your bull at the other and you get as good results or better without even having one of these rapacious, bad-tempered animals on the place. No danger, no nervous tension. Just prompt, efficient service.

Mrs. E.M. Black,
Butter-Fat, *February 1955.*

pounds of milk a year. Along the way, the BC Artificial Insemination Centre was established at Milner. The Chilliwack Artificial Insemination Centre had already been established. Now we had the makings of a perfect set-up to find the superior males and superior females from which we could get our milk. Herds on test for milk production were either provincial (the Dairy Herd Improvement Association) or federal (Record of Performance). Herds on test for milk production provide records which show animals' genetic potential by identifying the sire and the daughters from the mating from the records of many daughters on test. That bull can receive an accurate index into its genetic ability to produce milk and fat so the AI centres ended up with all superior sires.

When I retired 20 years ago, the average cow on test in the Fraser Valley was producing over 17,000 pounds of milk a year. This has been done through AI, and from frozen semen, and from all the other progress that was made by the other branches of agriculture. Better fertilizing, better methods, better corn, better breeding and all those things that happened in the rest of agriculture. I believe they're up over an average of 20,000 of milk a year now for each cow producing.

In 1913, over 6,000 lbs. milk was the average yearly production and herds on test. In 1947, the average amount of milk was 9,000 on test. In 1973, it was 13,000 pounds of milk a year. In 1993, it was 17,000 pounds of milk a year. This will give you an idea of what happened in the whole of agriculture soon after the Royal Commission on Milk. That started it all, in my opinion.

I made the statement at a meeting one time that they can go above 22,000 pounds of milk per cow, per year. It got a real rise out of the crowd. I don't want to give you the impression that it has just been AI and ROP testing of cattle that have made these changes possible. It was the whole thing. You see, we came back from overseas and most of the fellows that were in with us in the Department of Agriculture were post-war university graduates. Now all this whole group of people was out in the field — entomologists, horticulturalists, the whole works. These fellows were well informed and they were very ambitious to do a good job. It was truly, truly amazing. Veterinary service districts were established at Fort St. John, Dawson Creek, Smithers, the Okanagan, and other places. All this helped to get veterinarians established in the remoter areas of the province.

The BCAI Centre

Gordon Souter, 2000.

I started in AI in 1953 when I applied for a job as technician at the Chilliwack AI Centre and worked there for about three years. Then I went farming for awhile. Later, I worked at the BCAI Centre as a technician and through just being available to do whatever sort of different things there were to do — working with the bulls, attending meetings, speaking at meetings — I gradually worked my way up. I was always very much interested in the breeding and selection of animals, and so I got involved in sire selection in the coloured breeds then in beef sire selection. Later I did all of the sire selection, including the Holsteins.

True type drawings from the original paintings by Ross Butler approved by the Holstein-Friesian Association of Canada, showing the changes in the ideal animal. The cow in the top picture is a modern animal, while the one in the lower picture represents an older true type ideal. Type drawings and sculptures were developed in the early 1920s by the American Holstein Association to give farmers a visual goal for their breeding programs. Other breed associations developed their own true type drawings.

A Busload of Dairymen

Recently on a business trip that took them to Vancouver, SODICA General Manager, Everard Clarke, and President Ed Stickland took advantage of the opportunity to visit some Fraser Valley dairymen. They also made an extensive study of the BC Artificial Insemination Centre at Milner and attended the 75th anniversary of the Dominion Experimental Farm at Agassiz.

So impressed were they with what they saw that they resolved to discuss the matter with the board of directors and strongly recommended that Okanagan dairymen be given the opportunity to visit these two establishments early in the fall. The SODICA board of directors readily sanctioned the proposal and Bill Cameron, together with Des Hazlette and Jim Ryder of the Department of Agriculture, have now worked out details for a busload of Okanagan farmers to make a two-day trip to Milner and Agassiz on September 25 and 26.

The Cream Collector,
August 1962.

When the job as general manager became available in 1977, I applied for it, was hired, and did that for 20 years. Recently I've gone back to sire selection. I didn't have any formal training in university, but through reading and being interested and talking to a lot of people, I was able to learn quite a bit about the industry and about sire selection. That's how I got involved.

The Chilliwack AI Centre and the BCAI Centre were very much separate. At one time, there was an AI centre on Vancouver Island, one in the Okanagan, one in Chilliwack, and one here in Langley which originally was in Surrey. While they operated as individual cooperatives, they certainly didn't cooperate very much between one another. However, the one on Vancouver Island and the one in the Okanagan ran into financial difficulties very early, and both of those joined in with the BCAI Centre. The Chilliwack unit was more successful because it was right in the middle of the dairy industry at the time. It was in 1986 that they amalgamated with the BCAI Centre.

When I first started, we still used all fresh semen. Collection was in the morning, the phone calls from the farmers were taken by 10:00, and then we went out and did the insemination work. Dr. MacPherson, at the University of Guelph, started some work in frozen semen in the 1950s, but it wasn't until the late 1950s that the change to frozen semen happened in Ontario. I'm not sure when we changed, but it was probably the early 1960s. We used either homogenized milk as the buffer or egg yolk. It was a little bit odd that in Ontario they tended to use milk ex-

tender, and in BC we used egg yolk extender. I'm not sure why. Just to be different maybe. Early on, they were having major problems with conception rates. In about 1950, the BCAI Centre brought in Dr. Irwin to be general manager and to work on improving conception rates. He did a very good job of improving how the semen was handled and prepared, and that made quite a difference. Conception rates seem to have gone up to a certain level, and even with all the new technologies, they haven't changed that much. They just don't seem to go any higher than around 70 percent. They've been like that for 30 or 40 years.

Much of the work of the AI centre in the early years was educational. When I first started doing meetings, I didn't talk a lot about bulls. I learned (or tried to learn) quite a bit about reproduction and why cows repeated and didn't conceive early, and at most of our meetings with farmers, we'd talk about fertility and nutrition. While people wanted good bulls, and they felt that AI could give them better bulls, they were more concerned about getting cows in calf.

I think the two main reasons that AI got started and developed quite quickly were the danger of having dairy bulls around the farm and the transmission of disease. In the late 1940s and early 1950s, a number of people were killed on farms in Canada by dairy bulls. The other reason was transmission of disease by bulls. At that time, someone would have a bull and use him for a couple of years, then sell or trade him to another farmer. He would use him for a year or two and so on. Disease spread quite rapidly in many areas that way. The idea of being able to use the best bulls,

B.C. ARTIFICIAL INSEMINATION CENTRE

Milner, B. C.

Second Class Mail

Registration Number 1972

CANADA'S HIGHEST PRODUCING HERD

J. W. (Jim) Holt, Gerry and Jim Jr. with Part of Their High Producing Herd

In 1968 the Holstein herd of J. W. Holt and sons of Richmond, B.C. averaged 18497 - 700 - 305 - 2X.

This is the highest herd average for large herds ever attained in Canada. The Holt milking herd numbers about 50 head, all sired by B.C.A.I. Centre sires. The herd has been bred artificially for over 15 years and now totals about 120 head, including young stock.

Of the forty nine cows in the 1968 herd average, 17 produced over 20,000 pounds of milk in 305 days, and three two year olds produced 15,000 pounds of milk in 305 days. Segis Flo, a 7 year old daughter of Frasea Segis Vrouka Netherland, produced 23,845 - 908, Loretta, a 5 year old daughter of Gilmore Dekol Profile produced 24,978 - 898 and Sunny, a two year old daughter of Sunnyhome Aviator Successor, produced 16,367 - 635. The herd is off to a good start again and the Holts hope to top the 700 pounds of fat in 1969.

Lunch Time Crowd

These triplet heifer calves were born February 23rd at the farm of R. J. Barnes, Pitt Meadows, B.C. Their sire is Seiling Reflector (Ex) and their dam is Pitt Madcap Princess (GP) a daughter of Frasea Madcap Butterboy, a former B.C.A.I. Centre sire.

Princess has seven completed lactations that total 120,185 - 5157 - 4.29%.

The odds of triplet calves of the same sex being born are 20,000 to one. All three are doing fine.

Chilliwack Sire Available

KORWEST MARK, a son of Romandale Reflection Marquis, Excellent, is now available for sampling.

Mark's dam is Woodeyfields Susie Frasea, Good Plus, with six completed records, including her five year old record of 21,160 - 706 - 3.34% - 305. As a seven year old she produced 20,498,674 - 3.39% - 305. Her six lactations average M 180 - F 164% of B.C.A. She is a daughter of Frasea La Vata Sir Jewel, Very Good, a former B.C.A.I. Centre sire.

Mark is owned by the Chilliwack A.I. Centre and is being sampled under the joint proving program.

BCAI Centre Newsletter, 1968.

SODICA directors visit the BCAI Centre, 1962.

Gordon Souter, 2000.

Raising Champions

Professor John Berry, Department of Animal Husbandry, UBC, has often said "there is no magic formula for raising a champion cow but the study of lineage and genetics is a good step in that direction."

With the specialized farming of today, the raising of heifer calves has become an important part of farm management. Most efficient dairymen realize the importance of selecting heifer calves to increase production in their herd as compared to adding mediocre cows for a short term only.

NOCA Pictorial,
September 1968.

B·C·A·I NEWS

Published at Milner, B.C. by British Columbia Artificial Insemination Centre

VOL 2 NO. 4   DECEMBER, 1959

Talking It Over

Dean O. Shantz

It is with a great deal of pride and pleasure that I look back over the success that the members have enabled us to achieve in 1959. The present figures indicate that we will finish this year with 60,000 first services. This continued growth has enabled us to bring you service from the best sires obtainable with no increase in breeding fee.

In another section of the Newsletter will be found a brief itinerary of Max Brabant's (Assistant Mgr.) report on his attendance at an A.I. meeting in Ontario. It is this exchange of techniques and ideas that keep us abreast of latest research in A.I., thus assisting us in maintaining top efficiency in our program of supplying the best of semen from top pedigreed sires.

As this issue will reach you during the holiday season, I want to take this opportunity to remind you that there will be **No Service on Christmas Day or New Year's Day.** I know I have the co-operation of the members in our effort to have all the men home with their families on these two significant holidays during

Greetings

the year.

As 1959 closes, I hope we have been able to bring to you the service you have required and trust that our relationship will be as friendly

in 1960.

The Board of Directors, entire staff and myself join in wishing everyone a Very Merry Christmas and a Prosperous New Year.

BCAI Newsletter, 1959.

Program that the bulls available through AI really began to improve. Until we got into selecting young bulls and finding what we felt were the best pedigree cows that were in the breeder's herd and mating them to the best bulls we could find in the country, the bulls we had to market weren't all that great. Once we began the Young Sire Proving Program, we began to see real improvement. It really changed how dairymen looked at the information available on a bull's daughters. There were a lot of disappointments early on, too.

There has been a shift over the years from us sending out a fleet of technicians to a situation in which most of the AI work is "do it yourself" or owner insemination. In fact, in the prairie provinces, it's probably 95 or 96 percent "do it yourself" AI. It really boils down to economics. As herds got bigger, dairymen felt that if they were able to inseminate their own cows, they could reduce their costs. They also felt that they may have a bit of an advantage in improving conception rates by being able to inseminate their cows at whatever they felt was the best time. Studies show that basically there is no difference in conception rates between farmers breeding their own cows or having a technician do it. But most farmers feel it's a little more economical to breed their own cows. When I first started at the BCAI Centre, we had about 30 technicians.

so to speak, was there, but when you look at what was available at the time through AI as far as pedigree, it was no different than what farmers had in their own bulls on the farm.

It wasn't until we began the Young Sire Proving

University of British Columbia Department of Animal Husbandry dairy cattle judging team shows off trophies as the high team in Guernsey judging at the Pacific International Exhibition in Portland, Oregon, September 1947. Left to right: Dr. J.C. Berry (coach); Fred Hutchings, Roy Wilkinson, Howard Longfield, Robert Irwin.

Top-Notch Herds

Steves Dairy had a top-notch Holstein herd. That's where Les Gilmore got his start. I belong to the Dairy Herd Improvement Association and I remember about 15 years ago they had some of the old records back to 1932 and the production figures showed, roughly, 250 pounds of fat per year for a cow and now it's up to 1,000 pounds or better. In the old days production was measured by the amount of fat produced. Now it takes in amount of protein and solids, etc. But the production per cow has probably tripled.

A. Bates, 1994.

Steves family prize bull photographed outside Richmond United Church, circa 1940.

We've got about eight now. For years, we charged five or six dollars for service that included semen and three breedings. I think we were the first organization that brought in a system that charged a service fee for getting to the farm, and a fee to breed the cows. It didn't matter if it was the first service or the fourth. We got quite a bit of flak initially, but we got support from some of the good farm managers and eventually the system went over very well and was adopted by AI organizations throughout the country.

For us at the BCAI Centre, one of the biggest issues was trying to break into the global market as far as marketing semen. One of the first meetings that I went to was in Toronto at the Ontario Association of Animal Breeders. It was made up of just the Ontario AI organizations. That would have been probably in the early 1960s. I remember being at this meeting, and they were selling semen to Mexico, to South America, and to Europe. We were not sending any; yet our bulls were as good as theirs. I remember getting up at one of those meetings and saying, "Well, why couldn't we be part of the bigger picture?" Basically, they told me to go home. But we kept working away at it and eventually we were able to develop the organization which is now Semex. That was a very long struggle in getting accepted as an equal partner. Overall that's probably been the biggest thing for BCAI Centre, getting onto the world map. In BC, we have only 4 or 5 percent of the dairy cattle population, but we were sampling about 12 to 14 percent of the total number of bulls sampled in Canada. Out of those, our results were quite high. In other words, we were

getting a much higher number of bulls returned to a proof sheet than the other units. So we've done very well.

From a type standpoint, I think some of the biggest changes have been changes in the size of cattle. I think that udders have been probably the biggest improvement. Feet and legs would be the other. So, if you look at pictures of cows in the early part of the century, they basically were quite low set. They had very large udders with large teats and legs that had too much set. Type has changed dramatically: we've tried to breed cows with udders that are more symmetrical with teats that are easy to work with as far as using milking machines. As cows began to be kept on cement almost all their lives we've had to develop cows with a stronger foot, a deeper heel and a leg that's not too straight, but not too curved.

The shift away from the coloured breeds really began to happen after the war. I think two things happened in the 1950s. One was that people from the Netherlands came to Canada. There was a large number involved in buying and developing farms and they were used to black and white cows. But I think really the major change was the Clyne Commission. Justice Clyne brought in quotas, of course, and that meant that everybody was on an equal footing. Quotas were based on a pound of milk. It didn't matter if the pound of milk was 2 percent butterfat or 5 percent butterfat. So, at that time, of course, the Jersey people were producing 5 percent milk, but not nearly as much as the Holsteins were producing of 3 percent milk. The Holstein breeders got larger quotas.

Minutes from the first meeting of the Holstein-Friesian Association of British Columbia, New Westminster, 1908. BC HOLSTEIN MUSEUM COLLECTION

Advertisement, *Canadian Guernsey Breeders Journal,* **October 1954.** BOB IRWIN COLLECTION

Disease Control

The intradermal testing for tuberculosis was quite good. It was very sound. If they found a positive cow, that was it. But with Brucellosis Abortis, if a cow was heavy in calf or had just calved when we drew a blood sample from her, the test would come back suspicious. About a month later the test would be negative. There were big losses. Some people lost their whole herd. I used to work at Okalla Prison, for example, where they had a nice herd of Jerseys. When I diagnosed Brucellosis Abortis there, I said to the warden, "You better do something about this because if some of those prisoners find out that they picked up undulant fever while they were in Okalla, you'll be in deep trouble." The had to slaughter the whole herd.

Dr Joe Lomas, 2000.

Switching from Jersey to Guernsey
Dorothy Davie, 2000.

I married Murray Davie, and I came to the farm. We built a new house and we saw these cows out there — just a bunch of scrub cows. One day when Murray came home, I said, "If I've got to look at a bunch of scrub cows like that, I'll go stark raving mad!" He said, "Well, we're not going to have Jerseys and get into a hassle with your folks." And I said, "Fine, we won't have Holsteins then either" because his folks had Holsteins. He said, "I don't ever want to show cattle against a Paton." They had Ayrshires. And I said, "Well, what about Guernseys? Nobody around here's got a Guernsey cow that I know of." And he said, "Well, that's right." So the next morning he went over to Point Roberts and saw Lobi Thorstenson. Lobi had about eight Guernseys and he swore by them. He really liked them. So, he said, "Murray, I'll take you across to Skagit County and I'll show you cows you won't believe. Beautiful animals." The next morning, the two of them took off real early, came home real late, and Murray had bought three purebred Guernseys. And that's how we started.

Murray was away at meetings a lot. I kept track of the cattle. He was president of the Canadian Guernsey Breeders Association two times. They decided to hold a three-year-old futurity. That meant that you entered a calf this year and paid two dollars. There was no limit to how many you could enter, and then next year when she was a yearling, you could keep her entered or drop her. You entered her for a higher price each year, and then on the third year you'd have to take her to the

Royal Winter Fair to show her there. I entered five daughters in the first year and I kept them entered for the whole three years. Then we took the cattle and travelled 3,000 miles to Toronto to show them at the Royal Winter Fair, and we stood first, second, third, and seventh.

At that time, we were milking at least 60 Guernseys. That's how to breed purebred cattle. Get them better and better and better with each generation, see what's lacking this year and correct it for the next year. Build a herd like that. You just can't buy them.

After we were in the business for awhile, I was cutting the lawn one day to the barn and I shut the motor off and went in and walked down past all of them, and I said to Murray, "You know, these cattle are getting too beefy. They are straight across, and if anything, they slope forward to the front. You can't get milk from that kind of a cow. You got to get them up in front." I just felt like I wanted to go and grab them right at the top of their shoulders and pull up forward. You need to have it about six inches higher at the front than they are at the pin bone, or the tail. I just said, "We just got to get some new blood in here right away." So we knew Ben Frederichs, and we knew that he was going down to the east coast, and I said to him, "Ben, when you're back there, take a good look around the units, and see if you can find a herd sire that's eight inches higher in front than he is at the tail." Ben came home, and he said, "Dorothy, I found your bull." I said to him, "I hope it's Flash. I've been following him in the American Journal." We got 30 vials of Flash semen. That's where we got those futurity animals. ∎

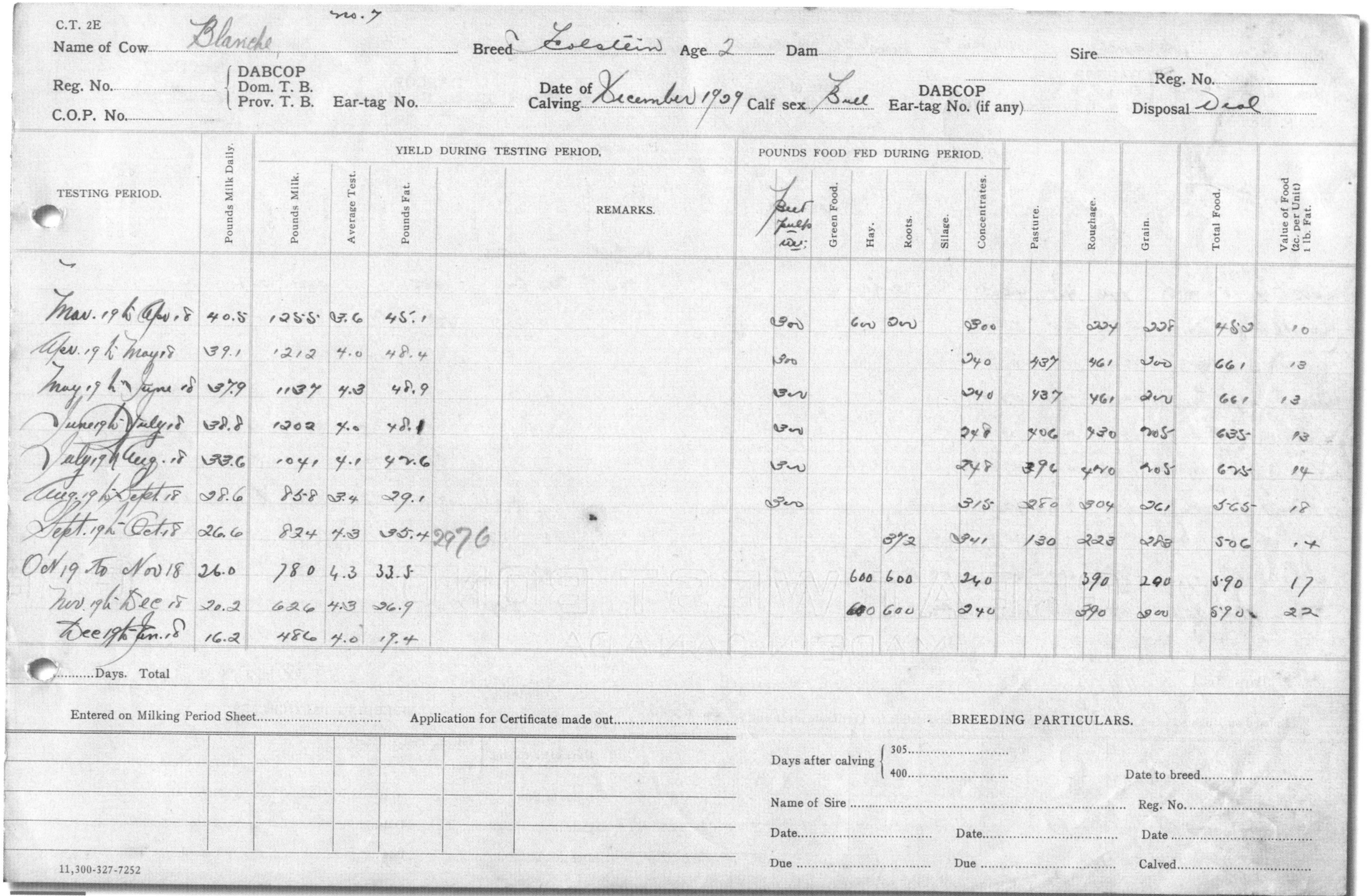

C.T. 2E

Name of Cow _Blanche_ no.7 Breed _Holstein_ Age _2_ Dam _____ Sire _____

Reg. No. _____ { DABCOP
Dom. T. B. _____
Prov. T. B. _____ Ear-tag No. _____ Date of Calving _December 1929_ Calf sex _Bull_ DABCOP Ear-tag No. (if any) _____ Reg. No. _____ Disposal _Deal_

C.O.P. No. _____

TESTING PERIOD.	Pounds Milk Daily.	YIELD DURING TESTING PERIOD.			REMARKS.	POUNDS FOOD FED DURING PERIOD.										Value of Food (2c. per Unit) 1 lb. Fat.
		Pounds Milk.	Average Test.	Pounds Fat.		Beet pulp raw	Green Food.	Hay.	Roots.	Silage.	Concentrates.	Pasture.	Roughage.	Grain.	Total Food.	
Mar.19 to Apr.18	40.5	1255	3.6	45.1		900	600 900				900		224	228	480	10
Apr.19 to May18	39.1	1212	4.0	48.4		900		240	437	461	300				661	13
May19 to June18	37.9	1137	4.3	48.9		300		240	437	461	200				661	13
June19 to July18	38.8	1202	4.0	48.1		300		248	406	430	205				635	13
July19 to Aug.18	33.6	1041	4.1	42.6		300		248	396	420	205				625	14
Aug.19 to Sept.18	28.6	858	3.4	29.1		300		315	280	304	261				565	18
Sept.19 to Oct.18	26.6	824	4.3	35.4 2976				372	341	130	223	283			506	.4
Oct.19 to Nov18	26.0	780	4.3	33.5			600	600		240			390	200	590	17
Nov.19 to Dec 18	20.2	626	4.3	26.9			600	600		240			390	200	590	22
Dec 19 to Jan.18	16.2	486	4.0	19.4												

_____ Days. Total

Entered on Milking Period Sheet _____ Application for Certificate made out _____

BREEDING PARTICULARS.

Days after calving { 305 _____
400 _____ Date to breed _____

Name of Sire _____ Reg. No. _____

Date _____ Date _____ Date _____

Due _____ Due _____ Calved _____

11,300-327-7252

Page from Richmond-Ladner Cow Testing Association Herd book, farm of Samuel Chorlton, 1929.

CHAPTER X

Processing

Bottling line at the FVMPA's Sardis plant, late 1940s.

For butter making from cream, a separator is the most effective and successful method of taking cream from the milk. It answers well to take from 16 to 25 percent fat from 100 lbs of milk.

The cream should not be kept for longer than 48 hours in summer before churning and the new cream cooled before adding to the cream crock, stirred well to mix and the whole kept at a temperature between 58 and 62° F.

When a ripening agent is added, the quantity should not exceed 2 ½ to 8 percent. Slightly better results are obtained by churning cream in the ripe stage and it should then be smooth, velvety and taste distinctly sour.

Good results from butter or cheese making will never be obtained from small holdings as a whole although individual cases will occur where fairly good butter will be produced, but it is only by handling fairly large quantities of milk and cream that consistent good quality is obtainable. Hence the advisability of small holders cooperating and starting a creamery or factory at a suitable centre where the product can be dealt with in a scientific manner by an expert.

Delta Times, *10 July 1909.*

Fifty years ago, a milkman on the streets of Vancouver offered a limited range of products delivered to the door. Today, some dairies offer an astounding range of products, many unrelated to the milk industry — juices, pizza shells, rice beverages. In the face of regional, national, and global competition, these dairies have left behind their allegiance solely to milk foods. David McMillan, CEO of Island Farms, for example, calls today's dairy a "consumer packaged goods business … And I say that because it is the consumer who dictates our security and our livelihood. We have to focus on the consumer."[1] The basic commitment of dairies to serve the needs of the customer has not changed over time; although in the past the means employed were very different.

In the era of the creamery, around the turn of the twentieth century, cream was the most valuable milk product and was an important part of the family economy in rural areas. It kept well, and travelled well, and could be transported over large distances to creameries where it could be made into butter, another product that kept well before refrigeration was common. In rural areas, skim milk in quantities greater than the needs of the family was dumped or was fed to pigs. In urban areas, or in those farming areas close to established centres, such as Grand Forks, Fernie, Creston, Kamloops, Nelson, Vernon, Kelowna, Delta, Richmond, and Saanich, the story was different. Farmers could bottle and sell milk on the fluid market as producer-vendors. According to historical researcher Ron Greene, thousands of small dairies have existed in the province since the early 1900s, most offering just fluid milk and some issuing tokens for dairy operations with as few as seven cows. In 1935, for example, Greene says 152 producer-vendors served the Victoria market. Before the 1920s, in many parts of the province, this raw, unpasteurized milk was not bottled, but was ladled from a five- or ten-gallon can on a horse-drawn wagon into jugs or bottles at the customer's door. Because there was no refrigeration, daily delivery was essential.

After the 1920s, milk and cream were delivered in round glass bottles with a cardboard cap in quarter pints, half pints, and pints for cream, and in half pints, pints, and quarts for milk. Milk was not homogenized, so the cream line was an important factor in comparing competing products; in fact, at some dairies, delivery men were instructed to tip the bottle slightly just prior to delivery so that the cream line would show to maximum advantage when the homeowner opened the door.

Milk itself was offered in a seemingly endless variety. Sometimes it was bottled under an individual farm's name and distributed as a specialty item. Sometimes it was fortified, in real or imaginary ways, by a variety of scientific means. Brooksbank Laboratories, for example, offered at a premium price not only certified milk, but also lactic acid milk and vitamin D fortified milk, long before vitamin D was a standard additive. Guernsey Breeders Dairy offered "golden" Guernsey milk because of its yellow colour. A Jersey Farms pamphlet of the 1950s offers standard milk, (3.5 percent butterfat) as an all purpose food, but urges consumers to buy homogenized milk (also 3.5 percent butterfat) because "every drop from top to bottom is consistently rich in cream content … it tastes richer, is more easily digested, and

Unloading cans at Northwestern Creamery, Victoria, 1940s.

has a pleasing texture for drinking."[2] Consumers were reminded that "homogenization does not impair the food value of the milk."[3] Other products included Jersey milk (4.5 percent butterfat), superior milk (9 percent butterfat), table cream (18 percent butterfat), whipping cream (32 percent butterfat), and buttermilk, "particularly delightful as a warm weather beverage."[4] Dairyland offered a chocolate flavoured dairy drink containing milk as well as cocoa, sugar, pure bourbon vanilla, and salt. It can be "enjoyed by all the family and is suggested for those who want to gain weight because it contains additional calories in the sugar and cocoa."[5]

Until many of the recommendations of the Clyne Commission were implemented, seasonal variation in milk production at farm were major problems for the industry. Summer production was double the production of the winter months, and manufacturing plants throughout the province were hard pressed to handle the summer surplus. The FVMPA purchased a condensary, which was processing Pacific evaporated milk, at Delair in 1924. This was a highly successful venture: not only did it allow the co-op to utilize seasonal overproduction, it was also a very popular product in its own right. Pacific evaporated milk was a household name in BC and as much a staple on the grocery store shelves as flour and sugar, particularly in isolated areas like the fishing and logging camps along the coast. It could be transported easily, didn't require refrigeration, had a long shelf life, and was versa-

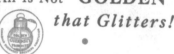

Promotional brochure, Guernsey Breeders Dairy. ACTON KILBY PAPERS, KILBY FARM MUSEUM

tile — reputed to be "wonderful" in coffee and just like fresh milk when diluted with equal parts of water. Pacific was also popular in urban areas. In the early 1950s, the Delta brand, containing half the butterfat of Pacific, was introduced to target the infant feeding market. Sales of these labels throughout western Canada peaked in 1970 at 1,200,000 cases per year and gradually tapered off as access to fresh milk improved throughout rural areas. Until the 1950s, the Borden Company operated a small condensary near Sardis for their sales of Borden evaporated milk in western Canada.

Independent dairies servicing the fluid milk market required a way to manage their historical requirement for a buffer to cushion demand through periods of low production and high demand, such as Christmastime.

Advertisement for Velvet ice cream, the Blue Boy restaurant, Vancouver.

Cottage Cheese

Cottage cheese was first produced for the FVMPA at Sardis in 1943. Walter Armitage was responsible for developing strict methods of production.

Walter Armitage was the type of person who we all had respect for because of the quality of product he was able to produce under, I would say, less then perfect conditions in the old Sardis plant. Even in those days, and in those conditions, he was able to continue to grow his cultures and inoculate the cans of milk for extending the starter cultures for the inoculation of the vats. He had to be a stickler for sanitation. Walter became known across Canada for his quality of cheese — he produced perhaps 50 percent of the total volume in Canada, which is significant. As a kid going into the department, I was terrified of him. I learned to respect him. I think we all did. He made a fine product.

Keith Miller, 2000.

The FVMPA's utility plant at Sardis handled milk surplus to the fluid requirements of all the dairies in the Lower Mainland as well as milk not required for the evaporated milk plant at Delair. This plant became the home of Fraser Valley butter — made only from fresh cream — and Sweetmilk skim milk powder. In the early years, Fraser Valley cheddar cheese was also produced here, but its production was abandoned after a few years because of the problem of whey disposal: a pig farm adjacent to the plant was able to take the whey, but local residents could not live with the resulting odour. During the Second World War, casein was produced at Sardis by treating skim milk with acid. Casein was a necessary component in glues required primarily for the construction of certain types of aircraft required for the war effort. Spray process skim milk powder for the bakery trade, roller process powder for the animal feed market, and Instant Pacific brand skim milk powder for the retail market were all processed and packaged at the Sardis plant in the late 1970s.

Okanagan producers managed their surplus in much the same way, by producing NOCA brand butter, reputed to be some of the finest in the country.

The consolidation of dairies began in 1931 when Fraser Valley Dairies plus 10 others formed Associated Dairies, a way for FVMPA producers to secure the retail trade in the Vancouver area. Just over 10 years later, the FVMPA purchased the remaining stock of Associated Dairies and renamed it Dairyland. By the middle of the 1960s, there were fewer than ten major companies in the provincial processing and distribution chain: Safeway, Dutch Dairies, Palm, Foremost, FVMPA,

Fraser Maid, SODICA, Island Farms, and Silverwood. Jersey Farms, which started in 1931, took in Guernsey Breeders, Glenburn Dairy, and Creamland Crescent in the late 1950s, and Richmond, National, Peters, and Hillside in the late 1960s. In 1966, Silverwood acquired Jersey Farms. Four years later, Silverwood sold its retail routes to Dairyland and in 1991 sold its remaining business and property to them as well.

Dairyland opened a new plant in Burnaby in 1962 and merged its Shannon Dairy operation with Dairyland in 1964. In 1968, it purchased the dairy in Kitimat, signing its first association shippers from outside the Fraser Valley. The same year, Comox Creamery was merged with the Dairyland operation. Improvements in the highway system throughout the province in the 1960s and early 1970s meant that processors in the Fraser Valley were able to service customer bases in the Okanagan and northern BC. Palm had plants in Nelson, Prince George, Vancouver, and Victoria. Dairyland created a network of branch operations in the larger communities — Prince Rupert, Prince George, Smithers, Kamloops, Kelowna, and others to meet customer needs and to help to keep them competitive with the Safeway and Kelly Douglas outlets. The Foremost plant became operational in 1967, serving the Kelly Douglas group of food outlets.

On Vancouver Island, Island Farms went through a similar period of consolidation. After incorporation in 1943, the company began an expansion program to serve customers up-island. Offices were opened in Duncan, Parksville, Qualicum, and Port Alberni. By 1954, however, the Port Alberni office — the last of the up-island outlets — was sold and business outside of

welcome
to Victoria's
modern
dairy!

ISLAND FARMS DAIRIES CO-OP ASSOC.

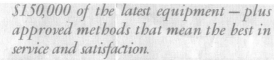

$150,000 of the latest equipment — plus approved methods that mean the best in service and satisfaction.

D. D. CHAPMAN
PRESIDENT

T. LINDSAY
MANAGER

A. WILLIAMS
SALES MANAGER
(MILK)

W. UDY
SALES MANAGER
(ICE CREAM)

FRESH MILK
from all the Island

From the Saanich Peninsula, from all Island Points as far north as Port Alberni comes the finest milk to be unloaded and received at Victoria's Modern Dairy.

FIRST STEP IN A
MODERN PROCESS

Pictured at the right is the Receiving Room where Raw Milk is received, weighed, graded and checked for flavor and started on its way via the cooler to the standardizer that equalizes the Butterfat content of all milk. Thence through stainless steel vats to the Pasteurizer. Also shown in operation is the Bottle Sterilizing Plant where repeated caustic baths, washing, rinsing and sterilizing sprays are automatically carried out to ensure complete sterilization.

OUTSTANDING PASTEURIZATION ROOM

Here is the Pasteurizer, shining clean new equipment where milk is heated to 161° and held for 16 seconds automatically, released completely pasteurized to the automatic filler and capper and delivered to the cooler bottled and complete with sanitary cellophane hood.

THE DAY'S SUPPLIES TO CITY HOMES
Ready for delivery . . . top-grade milk that has been processed for your table through Victoria's Modern Dairy.

ISLAND FARMS DAIRIES
Co-Operative Association

2220 BLANSHARD STREET   PHONE 2-315

Island Farms promotional brochure, 1950s.

the
Island Farms
story..

Island Farms was originally started in 1936 by a group of farmers in Saanich as a Private Company under the name of Registered Jersey Dairies. The business prospered and outgrew the limited capital which the owners had at their disposal, and in 1942 they sold out to a group of Vancouver businessmen who incorporated it under the name of Island Farms Limited. During the next two years the feeling grew among the farmers that they would like to regain control of their business, so on January 4th, 1944, the business was purchased and incorporated under the name of Island Farms Co-Operative Association, which has since been changed to Island Farms Dairies Co-Operative Association.

built —
owned and
operated by
the farmers of
Vancouver Island

Exterior of Island Farms Port Alberni plant, 1946.

Staff at Island Farms Port Alberni plant, 1949.

FRED MOCKFORD COLLECTION

Greater Victoria was handled by contractors. Pat Hamilton of De Clark Dairy at Ladysmith delivered in Nanaimo and as far south as Cobble Hill; Sidney Dairy covered north Saanich, Sidney, and central Saanich.

A new Island Farms plant was opened on Blanshard Street in Victoria in July of 1955. A new warehouse was opened three years later and land was bought adjacent to the warehouse site to accommodate future expansion. In the 1980s and 1990s, more land was purchased and a new ice cream plant and freezer storage warehouse were added.

In 1964, long-serving president of the Island Farms Dairies Cooperative Association, Willard Michell, stepped down. That year, John Pendray was elected president, a position he held for 30 years. In 1981, in an ambitious acquisition, Island Farms purchased Silverwood Dairies. In 1983, the plant was expanded, allowing Island Farms to expand its product range. Eight years later, in 1989, the Vancouver Island operation of Palm Dairies was acquired by Island Farms, leaving only two dairies, Island Farms and Royal Oak Dairy, in Victoria to compete with off-island dairies. Through the 1990s, Island Farms has worked to strengthen its regional position on Vancouver Island while at the same time worked to expand its market share off-island. It has installed state-of-the-art microfiltering machinery and has worked to develop a line of premium products, including micro-filtered milk, calcium fortified milk, and ice cream. At present, its president is George Aylard, the son of the first Island Farms Dairies Cooperative Association president, Arthur Aylard.

Quality Control

Quality control programs were introduced into dairy processing early. In 1927, the directors of the FVMPA recognized the need for a milk testing laboratory to monitor the quality of milk being received from shippers. Lyle Atkinson, a graduate of the University of British Columbia, was hired to develop a quality control program. As the program evolved, it included bi-weekly bacteriological testing of each members' milk shipment at each association plant, the employment of trained field staff to visit those members in need of assistance to maintain quality shipments, and the distribution of the in-house magazine, *Butter-Fat*, as a source of information relating to the production of clean, cold, milk. Other dairies had similar in-house magazines dedicated to the production of quality milk: SODICA published the *Cream Collector;* Island Farms published the *Island Farm News*.

Advertisement for Interior Creameries, *Alaska Highway News*, 17 August 1944. NORTH PEACE HISTORICAL SOCIETY, FORT ST. JOHN

Back of envelope advertising NOCA butter. LLOYD DUGGAN COLLECTION

Island Farms advertisement for Pure-Pak containers, *Daily Colonist*, 11 March 1962. GEORGE AYLARD COLLECTION

In 1933, George Okulitch, another UBC graduate, joined the FVMPA to assist Lyle Atkinson with the quality control program. In the same year, the association began an incentive program. To encourage compliance with the standards recommended by the laboratory staff, a penalty of 10 cents per pound of butterfat was levied on all milk not up to standard. Milk from each shipper was graded every two weeks as either Grade A, B, or C. Only Grade A milk was retained for the fluid market. Grade B and C milk were manufactured into animal feed powder at the FVMPA milk plant at Sardis. This program was unique in Canada because all milk, regardless of end use, was required to meet the fluid milk standard. Such a policy enhanced the quality of all manufactured products, particularly cottage cheese, and eventually yogurt. The quality control program was very successful. By the time that industry-wide regulations and standards were introduced as a result of the Clyne Commission, over 98 percent of the milk being received by the FVMPA qualified for the fluid market.

The Lucerne Milk division of Safeway, operating 17 dairy plants in the US, opened its first dairy plant in BC in 1955. The company brought to the Vancouver market a solid background in dairy processing and policies that included standards of performance for all phases of their fluid milk operation. With this background, Lucerne was able to encourage a number of large milk producers into becoming their suppliers. Although an American company, Lucerne quickly gained the loyalty of these producers by providing a field service program as well as a quality bonus program. Paying a bonus to milk shippers meeting its standards quickly allowed Lucerne to establish a high quality milk supply of sufficient volume to fully meet its needs.

At present, there are numerous processing plants operating in BC. Large regional or national markets are served by Agrifoods, the Lucerne Foods division of Canada Safeway, and Island Farms. Smaller regional dairies include Natrel Incorporated in Chilliwack, Blackwell Dairy Farm in Kamloops, D'Dutchmen Dairy in Sicamous, and the Bulkley Valley Milk Company in Smithers. Niche markets are served by many different processors. Ice cream is produced by Island Farms in Victoria, by Coast Mountain Dairy in Yarrow, by Blackwell Dairy in Kamloops, by D'Dutchmen Dairy in Sicamous, and by Birchwood Dairy in Abbotsford. Avalon Dairy in Vancouver and Royal Oak Dairy in Victoria continue to offer milk in glass bottles. Specialty product producers include Alamar Dairy in Delta, Fraser Meadow Farm in Chilliwack, the Saputo Group in Ladner, Flamingo Foods in Vancouver, Village Cheese in Armstrong, Ridgecrest Dairy in Mission, Punjab Milk Foods in Surrey, Meadowfresh Dairy Products in Burnaby, Gort's Gouda in Salmon Arm, and the Cowichan Cheese Company in Duncan. The organic market is served by Moon Struck Organic Cheese, Llewellyn Jerseys of Grand Forks, makers of Jerseyland cheese, and Greenfields Farms of Yarrow.

Milkman Unique

Frank Bradley, 2000.

There was little or no advertising as we know it in today's world. Some companies employed canvassers

Retirement dinner for Alan Burwash, manager of Jersey Farms. From left to right: George Okulitch, general manager of the FVMPA; Alan Burwash; George Fawcett, general manager of Palm Dairies, 1968.

FRANK BRADLEY COLLECTION

Retirement dinner for Alan Burwash, manager of Jersey Farms. Back row, from left to right: Dudley Burbidge, Jersey Farms; Frank Bradley, Jersey Farms; Herb Rhiel, Dairy Branch; Reg Cottingham, Island Farms; Norm Tupper, Dairyland; Svend Hansen, Jersey Farms; Frank Norton, Northwestern; Neil Gray, Dairyland; Ivor Fuller, Northwestern; Mac Sharpe, Palm; Don Guy, Silverwood; Jack Gray, Dairyland. Front row: Moffat Goepel, Dairyland; Roy Wilkinson, Dairy Branch; Grant Carlyle, Silverwood; George Okulitch, Dairyland; Alan Burwash, Jersey Farms; George Fawcett, Palm; Herb McArthur, Jersey Farms, 1968.

FRANK BRADLEY COLLECTION

Jersey Farms office and garden adjacent.

FRANK BRADLEY COLLECTION

Moose Milk

Bill Hewitt used to have wholesale route 102 down on False Creek flats. In those days, it was a residential area and there were an awful lot of corner stores and coffee shops and restaurants. He had been on that route for so long that he knew generations of customers. He knew the fathers and the grandfathers and the children. It was his Christmas tradition that he made up a batch of his famous — or infamous — moose milk from ice cream mix, milk, and I don't know what else. It was his secret recipe, but it did have large quantities of rum added to it. He would mix this batch up and he would put it in two-gallon cans and he would make his annual visit to his customers who would look forward to drinking it. Well, the customers each had one drink, but he had one with each customer.

Bob Cooper, 2000.

going door to door but mostly the search for customers was the responsibility of the delivery man. There was a variety of home delivery services available in those days other than milk. The postman, the ice man, the coal man, the bakery man, the wood man, the junk man, the Chinese produce man — all called regularly at the house. The milkman was unique, though, because he called every day of the week and most of his deliveries were made during the night. Remember, no electrical refrigeration and a much shorter shelf-life meant that it was a selling point if delivery could be made before breakfast. Virtually all of total milk sales at that time were made by home delivery.

Jersey Farms dairy was located in the 2200 block of West Broadway in Vancouver. Two blocks west was the Guernsey Breeders dairy and three blocks south was a branch of Associated Dairy, later to become Dairyland. These locations enabled the use of horse-drawn wagons to serve most of the areas on the west side of Vancouver. Trucks were used for delivery to the outlying areas. In case of heavy snowfall, sleighs were sometimes substituted for wagons.

The average milk route would serve 200 to 250 customers. They would place their empty bottles out on the porch before bedtime with money or tickets in the bottles. Most would have a standing order and would leave a note if they wished that changed. It was not uncommon to see two or three milk wagons serving customers in the same block at the same time. Competition was keen. Prices, as I recall, were fairly stable, so customer loyalty was achieved by good service and that was the job of the milkman.

He started work in the early hours of the morning and used a flashlight, or on the horse-drawn routes, some men used miners' lamps on their caps. I remember the acrid smell of the carbide when they were getting the lamps ready. The milkman's work day was a long one, often 10 to 12 hours with no overtime pay. This was occasioned partly by the slow speed of the horse and by the need to call back when customers were up and about to collect money owing or to sell tickets.

The milkman's job in some respects was similar to being in business for himself. He was responsible for the payment to the dairy for all the milk that he received for delivery, so he had to keep a record of transactions with each customer. He worked six weeks straight then got one week off. Prior to going on his week off, he had to total all the money owing and balance this with his account with the dairy. When he returned to work, the relief man had to balance the books and turn it back over to the regular man.

The last delivery by a horse-drawn wagon at Jersey Farms was in 1952. Most of the horses were heavy Clydesdales and the same horse would usually be kept on the same route. It was remarkable how well they knew the route and the customer stops — a real asset on a dark night or in the heavy fog. They were well treated. An oat bag was taken along and had to be tied on at a rest stop for the horse. The driver had to watch out when he was going down back lanes to make sure the horse didn't reach over the back fence and eat the customers' raspberry canes or other plants. During daylight hours, the children of customers often came out to feed the horse sugar lumps.

Dairy Section, David Spencer Store, Vancouver, circa 1915.
BCARS E09056

Garden beside Jersey Farms office in Vancouver.
FRANK BRADLEY COLLECTION

Competition

The competition between local dairies — Island Farms, Silverwood, and Palm Dairies — for business was intense, to say the least. Our salesmen would often return to the dairy to report that such-and-such a customer was about to change supplier because he had been offered a better deal or some free equipment or signage. Sometimes the competitor would deny that any such offer had been made. Letting a customer go was seldom acceptable, so something would be done to retain the business. Then another customer, maybe a friend of the one that had just been retained, would get wind of the new deal, and another round of negotiations would ensue. No one gave up a customer easily.

Des Thompson, 2000.

Accidents with wagons were rare, but occasionally the horse would cut in too close to a parked car. They didn't really need to be driven: they knew the route, and some were trained so that the driver could take a carrier filled with milk and cut through the back way to serve customers on the next street. The horse would come around and meet him there. They were pretty slow moving until it was time to go back to the dairy, then they perked up and managed a trot.

Dixie Cups
Bill Dinsmore, 1993.

I first started fooling around with dairies during the war years with Roy at Richmond Dairy on Hastings Street, kittycorner from the Waldorf. I was going to school and used to go down there at night with Roy and make dixie cups. At that time you couldn't buy sugar and we used to make ice cream from glucose. You couldn't buy dixie cup containers. We got water cups, filled them with ice cream, put a piece of waxed paper on top and packed them into wooden butter boxes. And then we put them into a little hardening room we had. We used to work there until 12:00 at night, lots of nights, filling dixie cups.

Condensing Milk
Bill Dinsmore, 1993.

In the early 1950s, we put in a condenser for condensing skim milk and were selling some of it to the Lucerne Ice Cream Plant and were using some of it in the Richmond Plant and we used to sell a little bit of it to Charlie Haslam up at Glenburn.

To make condensed milk, you take skim milk, heat it to about 150 degrees. When you put skim milk under vacuum (and that thing ran with about 20-21 inches of vacuum), it will boil at 140 degrees. The hot volatile gases will come off at the top — like the steam which goes into a water chamber that condenses it into water and puts it down the sewer. Now you've got condensed milk because you took the water out. They do the same thing when they are making powder — they carry the process further. They carry it on to a spray dryer — a roller dryer. But with the roller dryer, unless you had a good, experienced operator, the milk would burn on the drums and you would end up with brown streaks in the powder. Then, of course, it wasn't Class I powder and you ended up using it for pig feed.

Making Ice Cream at Northwestern Creamery
Ivor Fuller, 1999.

When I started with Northwestern Creamery in 1932 or 1933, all distribution was wholesale milk and ice cream. In those days ice cream was made with a batch freezer with ten gallons of mix at a time to produce somewhere in the neighbourhood, hopefully, of 20 gallons of ice cream, depending on the expertise of the operator. It was all packaged in one, two and a half, and five gallon containers. And at one point they packaged in two gallon containers as well. The containers were steel, returnable to the dairy. They were very expensive containers. Ice cream was also packaged in cardboard pints and quarts.

The ice cream was refrigerated by ice and salt in wooden tubs and all deliveries were made on that basis

Packaging ice cream, Northwestern Creamery, Victoria, 1940s. BCARS I-00981

Bottling milk, Palm Dairies, Victoria, 1947. BCARS I-01752

Soda Fountain, Northwestern Creamery, Victoria, 1944. BCARS I-00979

I had never seen a comptometer before coming to Island Farms. They were used to add up customer invoices and driver load sheets. By the time I left, they had been replaced by handheld computers and truck mounted printers used by the drivers.

Des Thompson, 2000.

except to restaurants and ice cream parlours and commercial outlets retailing ice cream. They had Waltham cabinets. These cabinets were cork insulated, usually of three square compartments, and each corner of the compartment held a triangle cylinder containing brine and refrigerated until it was frozen, and then it was carried with the ice cream truck. The cylinders were frozen at the plant in the Waltham cartridge room by direct expansion ammonia. This was from 1932 until 1937, right until we moved to the new plant on Yates Street.

Northwestern Creamery, Victoria
Frank Norton Jr., 1999.

Father formed Northwestern Creamery together with a partner in 1912. This was located on Cormorant Street, just below Douglas in what is now Centennial Square.

The Waltham ice cream cabinets used by Northwestern used to be called "slugs." During the winter they had to be cleaned with a wire brush. The plug was unsoldered and then they were filled if needed with brine (utetic solution). These were placed in a room that I helped build together with Dave Carlyle. This room contained rack on rack of ammonia coils and the slugs went right under the coils. It was a very expensive, inefficient system. In the summertime I would distribute the slugs and service them. In those days the 24th of May, picnic time, was always a big time. The Victoria Ice Company, owned by the Baker brothers, would come in every day with 300-pound blocks of ice over the sidewalk on Broad Street, up the alleyway, sliding

the 300 pound blocks past the old horizontal freezers into the high temperature room under the brine tank. We would have 10 to 15 blocks of ice all ready to come out to the ice crusher and be chopped up. The five-gallon steel cans of ice cream were packed in a wooden tub with ice and salt and, of course, were extremely heavy. The wooden tubs were made by Sweeney Cooperage in Vancouver.

Before we had the Waltham system we used to use ice and salt tubs. You would have to knock the wooden plugs out of the bottom of the tubs and drain the water out. Then you would put the fresh ice and salt mixture in and tap it all down and put the plug back in. This was just before the advent of electrical refrigeration. The Waltham cabinets were converted into electrical refrigeration by taking the sleeve out of the cabinet, wrapping it with copper tubing, and hooking up the compressor to it. This was around 1937 or 1938.

Palm Dairies, Victoria
Mac Sharpe, 1993.

I started with Palm Dairies in Regina, a completely owned subsidiary of P. Burns. I worked there from 1936 to 1945 in the office. I started at $12 per week and after a year, it was raised to $12.50. I had a number of positions at Palm in Calgary, Sudbury, Winnipeg, and Moose Jaw. In 1957, I was transferred to Victoria and was in charge of the branch until I retired in 1981.

The Victoria branch began when the company bought the assets of the Vancouver Island Milk Producers' Association in 1928 and changed the name to

FVMPA Laboratory at 8th Avenue plant. Lyle Atkinson is at the microscope and George Okulitch is standing at right, 1934.

The BC Federation of Agriculture was instrumental in having an amendment passed to the provincial *Creameries and Dairies Regulation Act* to prevent the use of vegetable oils in manufactured dairy products except margarine and to make it impossible for any dairy in BC to sell reconstituted milk without a permit.

Minutes of BC Federation of Agriculture Directors Meeting, January 27, 1953.

Palm. When I came to the branch, they did milk processing, ice cream, cottage cheese — everything but butter. We had our own shippers and in 1959 we moved from cans to bulk tank pick up. At the same time, we bought Sunnybrae Dairy in Duncan. We carried on with that operation until around 1969 when we bought Baby's Own Dairy in Naniamo. That was the extent of our operation. We had distributing points in Duncan and Nanaimo and Courtenay. When I first came to Victoria, we had an ice cream operation in Nanaimo. At that time, they were starting to centralize by making all the novelties at one plant, and cutting down on the variety. George Fawcett was manager in Vancouver. Palm Dairies operated in Nelson for a long time. Prince George was purchased sometime in the 1960s and Bert Abercrombie was manager there. We also had a branch in Trail at one time. We were never successful in the Okanagan. We used to ship into that area, but never really got a good toe-hold. Everard Clarke was a pretty strong man to compete against.

I was manager in Victoria for 24 years. One of the things we got in was the five-day week. We worked no Wednesdays, and no Sundays. Men knew they would get those days off every week without fail. It seemed to be a desirable schedule for them. We were splitting market share with Island Farms and Northwestern. Lots of milk came to Vancouver Island — packaged and in bulk — from Lucerne and the FVMPA. After I left, Jerry Hellier was made manager, and he carried on for two or three years and was sent to Calgary. Bill Keck took over and the company was bought by Island Farms and closed up entirely. It was quite a surprise.

The Cream Line
Marg Savage, 1993.

The Savage farm always had Jersey cows. When they shipped to Jersey Farms the big thing was cream line. We used to tip the bottles so the cream line was evident. After the war, a lot of breeders realized that a volume of milk was needed and people started to produce for volume. The Holstein cow became popular because of her high production. And the way of life changed: people want less fat now than before.

Even 20 years ago, it was common for the field man of the dairy (Lucerne or whomever), to drop in at the farm. Now, there's no personal contact with the dairy plant. I think co-op members felt they were closer to the dairy plant. As the years went by, a lot of the breeders became less involved with the dairies. As the little dairies fell by the wayside there wasn't that personal touch any more where the producer could really relate to, or get involved with, what was happening with the dairy. The breeders produced the milk and let the dairy plants sell it. You got right away from the personal touch.

A conversation about starting out
Frank Bradley, Bill Osborne, Jeff Harbottle, 1993.

Frank Bradley: It is amazing that so many started out during the Depression years, with just a truck and six quarts of milk in the back. There was a whole lot of them: Guernsey Breeders, Charlie Haslam out in Glenburn, Richmond Dairy, Creamland Cresent, Empress Dairy.

Bill Osborne: In those days they were localized. They had horses and would go to the limit of where a horse could go ... Every district had a little dairy.

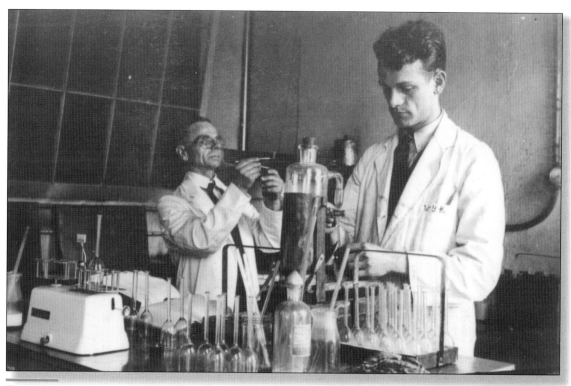

FRASER VALLEY BUTTER

Discriminating Housewives Prefer "Fraser Valley" Butter . . . .

Because it is the Best!

Because of its unvarying freshness.

Because of its Sweet Cream flavor and delicious taste and

Because "Fraser Valley" Butter is a truly British Columbia product, creating employment for British Columbia farmers and workers.

These reasons are given in hundreds of signed testimonials received from discerning buyers of butter, who realize that in "Fraser Valley" Butter there is a superior quality and unequalled value.

How much more then should this, your own product, be valued by the members of the F. V. M. P. A., whose livelihood and prosperity depend upon its sale?

Ad for Fraser Valley Butter, 1933.   BUTTER-FAT, 1933

Freddie Washington (left) and Clifford Kelly (right) testing milk at the FVMPA's 8th Avenue plant, late 1920s or early 1930s.   DAIRYWORLD COLLECTION

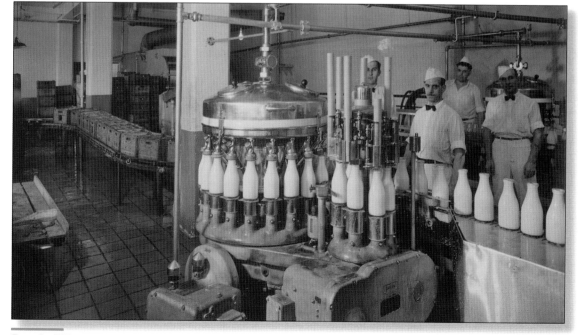

Bottle Fillers, Associated Dairies, 1930s   DAIRYWORLD COLLECTION

Ken Savage

Ken Savage was Dairy Commissioner. After he left, he became chairman of the Canadian Dairy Commission and after that, he went to the IDF — the International Dairy Federation — in Brussels where he was chairman for two-five year terms. He was a very prominent individual.

Neil Gray, 2000.

Bob McArthur, plant manager of Lucerne Milk at 995 Mainland Street in Vancouver, leaving for job in Bellevue, Washington plant, 1955. Front row, left to right: George Hunter, Dorothy Gilbert, Bob McArthur, Pete LeMarrec, Hazel Chinn, Max Rohweder, George Anderson, Hugh White. Back Row: Jack Kuiper, Frank Kadla, Hans Morthorst, Bill Heath, Bob Irwin, Gordon Mohl.

BOB IRWIN COLLECTION

Jeff Harbottle: And there wasn't the government interference then as there is now. You could just go down and get a licence and do anything.

Beautiful White Tails

Dorothy Davie, 2000.

> We started shipping milk to Safeway. They paid a premium at that time. Guernsey milk is different, you know, a different colour. Even the skim milk is a better colour than I've ever seen in a Jersey or Holstein. Either of those two tends to be a little blue, and Guernsey milk is never like that. The skim milk is a bone colour. It has extra keratin. At the end of every Guernsey cow's tail, if you part that white hair, there's a little orange coloured ball of keratin. The Guernsey cows have beautiful white tails. We had one friend who lived in Tsawassen, and he said one of his greatest pleasures was to drive to work at nine o'clock in the morning and see the Del Eden herd of cows fanning out over the green pasture. It was a sight to see.

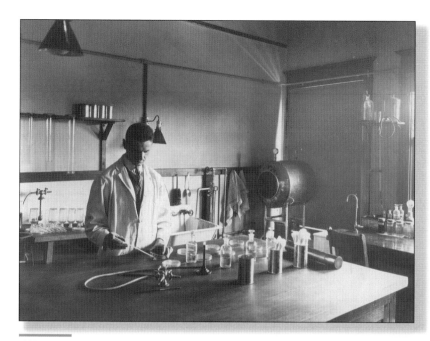

Clifford Kelly, first bacteriologist at the FVMPA, 1924. DAIRYWORLD COLLECTION

Boxing NOCA cottage cheese. LLOYD DUGGAN COLLECTION

Wayne Skebinski, foreman, in control room of Lucerne's new Burnaby plant, early 1970s.
BOB IRWIN COLLECTION

Lucerne Milk Company

Bob Irwin, 2000.

In 1955, I left Jersey Farms to accept a position with Lucerne Milk Company. Lucerne Milk, with its head office in Oakland, California, was part of the dairy division of supply company operations of Safeway Stores Incorporated. The dairy division operated fluid milk, cottage cheese, ice cream, evaporated milk, powdered skim milk, and cheese cutting and packaging operations in 17 plants across the United States and Canada. Three of the plants, located in Winnipeg, Edmonton, and Vancouver, supplied Canada Safeway Stores.

Developments during and after the Clyne Commission led to the granting of a fluid milk plant licence to the Lucerne Milk division of Canada Safeway. A milk and cottage cheese plant was installed in the existing building at 995 Mainland Street in Vancouver on property shared by Mcdonald's Consolidated and Edwards Coffee.

Milk processing and packaging started in January 1955. Lucerne became the exclusive supplier to Lower Mainland Safeway stores. Eventually, Lucerne was distributing from Vancouver to all Safeway stores in BC. In certain areas, local suppliers were allowed "split stocking" with Lucerne. Lucerne Ice Cream division operated a separate plant near Granville Street and 2nd Avenue in Vancouver.

The establishment of a store differential of two cents per quart was an incentive for consumers to switch from daily home delivery to store purchasing of milk and dairy products. All Lucerne products were expiry dated and sold with a "three times the value" replacement policy. Fluid products were packaged, not in bottles, but only in hot-wax coated Pure-Pak cartons.

The BC Milk Board assigned 18 milk producers to Lucerne, including six on Sumas Prairie that had been shipping to Palm Dairies since 1953. Each farm's milk house was equipped with a refrigerated cooling and storage bulk milk tank sized to hold at least four milkings. Tanker truck pick-ups were scheduled on an every-other-day plan in a 1500 gallon tank truck operated by R and R Trucking — "Mac" McKinnon of Abbotsford. The plant was not equipped to receive milk in ten-gallon cans.

From 1968 to 1970, I was directly involved in the planning, construction, equipping, and start-up of our new Burnaby plant, a $3.8 million fluid milk and cottage cheese plant with delicatessen kitchen. The plant commenced operations in October of 1970. In 1972, I went to work for the FVMPA, first as director of laboratories, and later as manager of laboratory and farm services. I retired in 1991.

Opening New Plants

Maury Roberts, 1993.

One day in 1960, I was working in my machine shop in downtown Vancouver and looked up and there was an old friend, Neil Gray. At that time, he was employed by Creamery Package Manufacturing Company, a manufacturer of food, dairy, and refrigeration equipment. He asked me to go with him to the Vancouver Hotel to meet his boss from Toronto. So down we went, me in my coveralls, to meet Harry Nellist. Neil had

Control panel for milk pasteurization and distribution lines to storage and packaging machines at Dairyland's new Burnaby plant, 1964. Grant Larkin, plant superintendent, is at the controls.

Christmas Cakes

In 1948 or 1949 Dairyland decided it would sell ice cream cakes to the Christmas market. First of all, we had to develop a cake. We developed a sort of Christmas fruit-and-rum-flavoured ice cream. We ended up buying every cake pan in Vancouver of one particular size. The cake was turned out of the mould and then decorated with whipping cream in a fancy design on top. Marketing went ahead to try and sell this thing through home delivery. As Christmas got nearer and nearer, the orders got greater. We were concerned because we only had so many tins to make the cakes in. It snowballed to the point that we were going well into the night making ice cream to freeze in the tins and then dumping it out and decorating it in the morning. Three or four days before Christmas, there were literally thousands of orders to fill with a limited number of tins and personnel. It was a very difficult time. We worked until 12:30 on Christmas eve and then came back at 6:00 on Christmas day and dumped them out and started decorating again because they had to go on the trucks. I don't remember a single, solitary thing that provoked more difficulty, more problems, more complaints, than that ice cream cake.

Grant Larkin, 2000.

accepted a position as manager of Shannon Dairy and had suggested to Harry that I should take over the BC and Alberta territory for Creamery Package. So the upshot of that was I did, and remained with them until 1966. I had just joined Creamery Package when the concept of three more plants for Canada was initiated: Foremost Foods, one plant; Lucerne milk plant in Burnaby; and Lucerne milk plant in Winnipeg. This Foremost plant was designed and developed by Creamery Package through George Haxton, Western Regional engineer for Creamery Package, out of California. He knew how to put a milk plant together. The Safeway plants were partly done through George Haxton and the Safeway people, a concept of the new generation of milk plants. Because I was there once the plants were initiated, the contractors and designers — all the Americans — went back south and I had to put these three plants together. That was the highlight of my life in learning because I either had to put the plants together successfully or fade into a hole in the ground. I burned the midnight oil.

I was brought up in Calgary until I was 12 years old. In those days my father travelled all over Western Canada for Walker Wallace. We left Calgary to come to Vancouver in 1926. When I was old enough, on Saturdays my dad used to take me on some of his calls to the dairies — United Dairies, Union Milk, the Coldcutts Dairy, Model Dairy. So that was how my travels in the dairy industry started. We spent a couple of summers — 1924 and 1925 — in Vancouver and rented a summer cottage at Jericho Beach. My mother liked Vancouver so much the family moved.

I went to Vancouver College. As a kid I was always interested in automobiles from the mechanical side. I went to the public library and found out how an automobile works and I was always fiddling with cars and repairing the neighbours' cars when I was just a kid. I worked around the service stations while I was going to high school and after the war I opened up my own business specializing in brakes and motor tune-ups. Previous to that time I went on a number of trips with my dad installing equipment. We installed the HTST (High Temperature Short Time) system over at Northwestern Creameries on Yates Street in Victoria. Francis Norton was there at the time. I did all the work and he did all the directing. I did help Dad out quite a bit — putting in a filler at Jersey Farms. Then my dad got involved with the foil over-cap and I got deeply involved in that one, making sure those things worked. I helped my dad put the Short Time equipment in Jersey Farms. The Short Time equipment was invented by Dr. Sleekman of APV in England. That was in the 1920s. I don't think Dr. Sleekman was the original inventor of the plate machine but he was the original inventor of the plate machine of the sanitary type for the dairy industry. Mr. Walker (from Walker Wallace Inc.), around the early 1930s, went over to APV in England and from Dr. Sleekman he received the licence to distribute the plate machine in North America — it was the first plate machine in North America. By that time APV in England had developed the Short Time and he brought the plate machine and the Short Time over to Canada and the United States.

The Short Time was an instant success. I say that

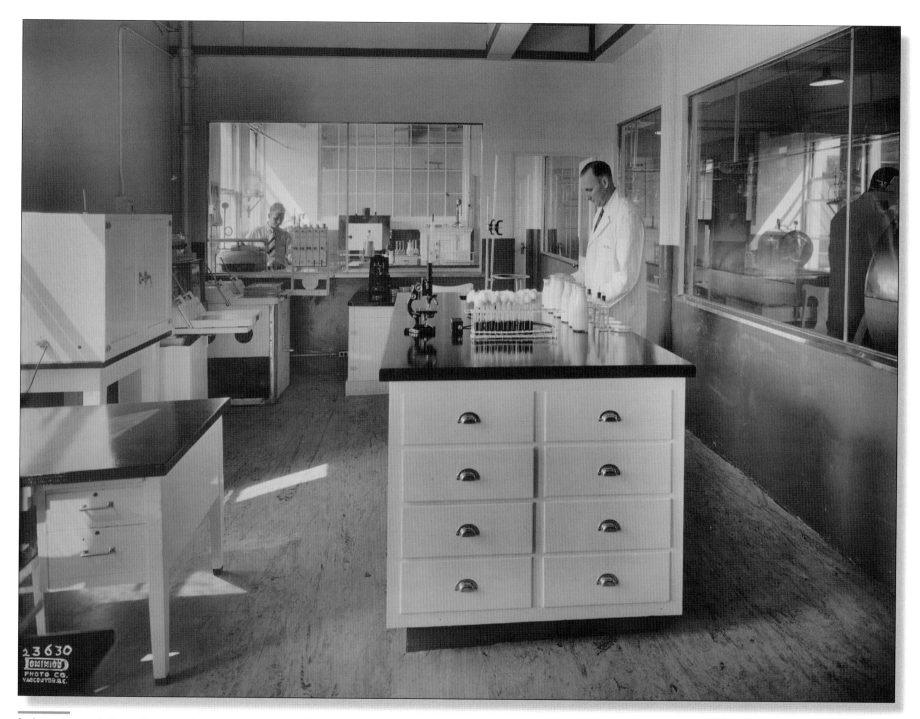

Laboratory of Associated Dairies with George Martin working (right), 1930s.

Responsibilities

We all have a responsibility, you and I, to utilize the facilities known as the Fraser Valley Milk Producers' Cooperative Association in such a way that we can pass them to the next generation, as they were to us, ready and capable of serving.

Peter Friesen, "President's Report to FVMPCA," 1986.

with reservation. It was not very kind to the cream line and that was the big drawback. I remember over at Northwestern Creamery when we got that piece of equipment working, we made arrangements on Monday morning to have the plant people start it up and we went in there at 7:00 a.m. to start it up and the old system of vat pasteurizing would be out of commission. But the plant people started up the old system at 6:00 a.m. It took us 10 days before we could get the plant people to finally start up the Short Time System. The CIP (Clean In Place) was also a big advance. It came with the Short Time. It hasn't changed very much from today.

Systems handling was another change. That changed the methods of handling in the processing plants too. As the plants got larger, the business of walking and carrying things became too time consuming so the conveyor system had to start improving.

Dairyland

Bill Ramsell, 2000.

I think that the biggest change came after the war when we went to homogenizing all the milk. Until then, most people looked at the cream line. That was their judge of how good their milk was.

When we phased out bottles and started with cartons on retail, I remember we had a route that went

Opening of new Dairyland plant in Abbotsford. From left to right: general contractor, Randolph Allan; Dairyland General Manager Neil Gray; FVMPA President Gordon Park; and Matsqui Mayor Harry DeJong, 1980.
NEIL GRAY COLLECTION

out to Port Moody and up the valley to Anmore. So we started it on the route and I remember I met the driver one morning about 7:30 and I was wondering how people were taking it, because somebody said that they

British Columbia Department of Agriculture.
(Dairy Branch.)

REPORT ON MILK PLANTS, CHEESE-FACTORIES, AND ICE-CREAM PLANTS.
(Confidential.)

Name of firm *Mc Crindles Dairy*

Address *Cranbrook*

Wholesale or retail *retail*

Products manufactured or processed *Pasteurized milk & Cream*

Trade-names of products _____

No. of milk shippers *own Milk* No. of table-cream shippers ____

No. samples tested for B.F. _____ Methylene blue _____

Resazurin _____ Sediment _____ Bacteria count _____

B. coli. (pos. or neg.) _____ Phosphatase (pos. or neg.) _____

No. of milk-cans inspected _____ No. declared unfit for use _____

Pounds milk received per day from farmers _____

Pounds milk received per day from other sources _____

Pounds table-cream received per day _____

Pounds cheese made since January 1st _____

Pounds cheese made same period last year _____

Gallons ice-cream made since January 1st _____

Gallons ice-cream made same period last year _____

INSPECTOR'S GENERAL REPORT.

Mr. Mc Crindle's dairy was in the usual spotless condition. He has installed a pasteurizer and new cooler and a 3 compartment bottle washing machine. He has enlarged his milk storage room and boiler room and has installed 2 boilers of 2.9 H.P. each. Thus obviating the necessity of a boiler certificate.

Mr. Mc Crindle hopes to retire soon and turn this business over to Cecil Morrison and Clarence Barrett. one of whom hopes to attend our short course.

Date of visit *Sept 20/47*

Geo. Patchett

Inspector under "Creameries and Dairies Regulation Act."

20 bks.-1144-3755

Pacific Evap. Milk Exports from B.C. Foreign

Central America	Quantity	Value 1935		Quantity 1936	Value
Central America				31 Tons	$ 6014·00
China	8300 Cases	$24,817·00	2.999 pr cs	15449 Cases	$ 53763·00
Egypt	5200 Cases	15288·00	2.94 pr cs	6600 Cases	$ 19800·00
Various European				4 Tons	$ 500·00
Fiji Islands	211 Cases	681·00	3.22 pr cs	388 Cases	$ 1405·00
Gibraltar	750 Cases	1980·00	2.64 pr cs	600 Cases	2172·00
Japan	170 Cases	549·00	$3.20 pr cs		
Malta	2490 Cases	6929·00	$2.84 pr cs	780 Cases	2824·00
Mesopotamia	600 Cases	1938·00	$3.26 pr cs	400 Cases	1448·00
New Zealand				50 Cases	181·00
Palestine	51 Cases	164·00	$3.20 pr cs		
Panama	800 Cases	2184·00	$2.73 pr cs	1650 Cases	7689·00
Peru	200 Cases	590·00	$2.95 pr cs		
Phillipines Isl.				200 Cases	724·00
"				72 Tons	10872·00
South America				250 Cases	1735·00
Strait Settlements	6880 Cases	20089·00	$2.93 pr cs	5588 Cases	17490·00
United Kingdom				1225 Cases	4055·00
"	75142 cs	293053·00			
"				1188 Tons	163706·00
West Indies	150 cs	1063·00		125 Cases	876·00

305

Home Delivery

were not very happy about it. I went up to one lady and she was elderly, and I said, " I'm just checking, how do you like these cartons?" "Oh," she said to me, "It doesn't make any difference, because I just wait for the milkman to come and I take the carton and I pour it into my bottle."

I can recall when we introduced yogurt, it was just straight yogurt. We had a training program at the dairy and we used to try and get the fellows to taste it, and they just wouldn't touch it. And they were supposed to be selling it. There was very little enthusiasm for it until we got into the flavoured yogurt and then all of a sudden it took off. We used to put people in the stores demonstrating. It was a real problem getting good demonstrators because they meant so much to you. There was one who later worked with Woodwards, Mona Brun. She got her start in the demonstration business through her efforts with Dairyland.

I think at one time we counted about 36 home farms that were selling their milk directly in Vancouver and their big idea was that it was non-pasteurized. One of the biggest arrangements was Grauers' farm. It was a real beautiful spot. You could go out there any time of the day or at night and it was spotless, and they produced unpasteurized milk; yet Grauer was one of the first ones to put in a pasteurizing plant. If there was anybody that had a sale for unpasteurized milk that was him.

But I think the biggest kick we got was when we went into adding vitamin D to the milk. Some people swore they could taste fish oil. I got the job of going around with a sample of vitamin D in a little bottle so

they could see it, smell it, taste it.

I can recall when we had the milk inspector. In Vancouver, he had a Ford car and he would go out in the morning and he would pick up bottles of milk from the wagons. He would sign a receipt for it. Then the driver went in and he turned that receipt into the dairy to say the samples had gone. There was no refrigeration in that car. In the hot summer, what those samples would be like when he got back and tested them at the police laboratory! That's where it was in the early days. In Point Grey, the municipality had its own milk inspector and they had a man who rode a street car with a little satchel that he carried his samples of milk in. So he'd take a sample from me say at 7:00 in the morning, and he might get it to the police lab at about 4:00 in the afternoon by streetcar. I would think a million-count bacteria would be a very common thing for them to find.

And it was interesting in the days of the Associated Dairies. Fraser Valley Milk Producers was on one side. The side where the milk from the farms came in, that was called the Fraser Valley side. Where it was bottled, pasteurized, and so on, was the Associated Dairies side. When it went through the pipe, it became Associated. And both sides took tests. George Martin was a milk inspector in Vancouver. When he was in our side, he used to swear that the Fraser Valley were cheating the Associated Dairies with their tests all the time. The Fraser Valley said that he used to cheat on all the tests that went on in our side, and so it went. There was that split right down the middle, and after we opened our new plant in Burnaby in 1964, that split sort of disappeared.

NEW!

Odermatt's Dairy Ltd. has just completed installation of the most modern milk processing and packaging equipment available.

NEW PUR-PAK MILK CONTAINERS

The modern look in milk cartons —

30 per cent space-saving in your refrigerator,

New easy-pour spout — no spilling, no dripping. Empties completely.

NEW VITAMIN D ENRICHED MILK

The "sunshine" vitamin added to our fresh whole milk — better for you . . . better for your children.

We are proud to be the first and only dairy in the entire Peace River area to offer these modern advances in milk handling

. . . and AT NO EXTRA COST TO YOU!

Always buy CHALET BRAND Products
from your progressive local dairy.

ODERMATT'S DAIRY
LTD.

PHONE 315   FORT ST. JOHN

Life is Short - -
Why take a chance?

•

Play Safe - Be Wise - Have Pasteurized

•

Dear Sir or Madam:

I think you will agree the above heading is a safe one to follow. For instance, in preparing food of any kind you see that it is properly prepared and you take no chances.

•

OUR BUSINESS IS MILK—which arrives at our plant direct from the farm, is graded and treated by a practical experience of thirty years. Our system of treatment—hot water method—(the way your doctor would like you to do it) gives our product that smooth and tasty flavor that all lovers of good milk enjoy. We carry no side lines. We stick to milk. We are an entirely independent firm and have stood the test of time and expect to serve you a long time yet.

The Avalon invites you to try its "Standard," which is good milk, Guernsey and Jersey.

Avalon Dairy

Phone Carleton 198   43rd Avenue and Wales Road

Avalon Dairy ad, undated.   ACTON KILBY COLLECTION, KILBY FARM MUSEUM

Advertisement for Odermatt Dairy, *Alaska Highway News,* **15 September 1960.**   NORTH PEACE HISTORICAL SOCIETY, FORT ST. JOHN

Changing Over

Walter Sorenson, owner of Nanaimo Dairy, said his plant went "Pure-Pak" overnight, which might be considered a rash move ... However, he summarily gained a nice 20 percent in the first month! Sorenson says, offhandedly, "I was quite amazed."

From "Ambitious, Beautiful Vancouver," in Pure-Pak News Pictorial, *1960.*

At one time we used to try to check on sales. As a matter of fact, we had a deal with the post office. We used to take our routes and tie them in with the postal routes and we could tell fairly accurately the percentage of our business. It was a good scheme. I remember when they moved the returned soldiers from the old Hotel Vancouver to a place up on the Heights, just around Hastings Park. They had all those houses built and they didn't turn the water on. One of our drivers got a key that could turn the city water on. So everybody was chasing him trying to get their water turned on. We did very well in there. It didn't hurt at all. But little things like that made a difference. The drivers got to know everybody and a lot of people got to know them and they were part of the community. I think that sort of eased off at the end as we got into these bigger routes and reduced deliveries. I remember one lady calling up and she was mad. She said her baby hadn't said a word and they were trying to teach it to say "DaDa." When the milkman came, she opened the door and the kid said "DaDa." And she was very annoyed. I don't know how you blame the milkman for that.

One interesting thing I recall years ago, was that all the offices in the Associated Dairies and the Fraser Valley — all the offices on 8th Avenue were pretty well taken up by executive people. Everybody had a spittoon and I always used to say that you haven't amounted to anything if you haven't got a spittoon in your office! It's sure changed. The janitors took care of that. As soon as anybody got out and they were finished, they emptied the spittoon and threw it away. I remember when Dougal MacDonald, the marketing manager for Pacific Milk, retired, they took his spittoon and had it all polished up and planted a geranium in it and presented it to him as he left.

Armstrong Cheese

Adrian Shrauwen, 2000.

We started in Kamloops with Dutch Dairy. Dad started small. When I was 18, I went overseas, and Dad converted everything into beef cattle. Then after the war, when I came back from overseas, we couldn't get enough milk out of Kamloops to serve our customers. We also had Valley Dairies. Ernie Carr of the Milk Board assured us we'd get milk, so that's the way it went. We took over Armstrong Cheese in 1961. The plant was not doing well and in the end they went into receivership. We were asked to make an offer on the plant. They called us and they called Everard Clarke to see who was going to get the best deal. In the end, we got it. Everything we touched was rusted. We couldn't start up. Nothing worked: it wasn't broken, it was rusted. We changed everything — it cost thousands and thousands of dollars. Once we got the plant going, we found out that our cheese production was not keeping up to our sales, so that's when we found another cheese plant. It was not originally a cheese plant, it was butter and fluid milk, a very small plant in Bashaw, Alberta. Later, we sold both plants to Dairyland.

Plant Inspection

Earl Jenstad, 2000.

As a young boy I gained experience on my grandparents' raw milk dairy farm in the Creston Valley. There I first

Women working on packaging line for **NOCA**, likely in Salmon Arm.

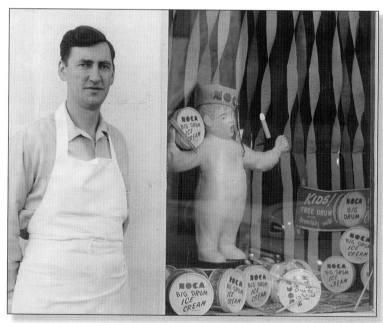

John Gluck displays **NOCA** drums of ice cream in the window of his Penticton Confectionery store.

NOCA's new milk cartons roll off the line.

In 1964 the Dairy Branch assumed responsibility for lab testing milk samples from every shipper. This responsibility was transferred from the Department of Health.

Roy Wilkinson, 2000.

learned sanitation procedures and the bottling of milk. I also learned how to milk cows, first by hand, then using milking machines. I never lost interest in dairying and on April 1, 1969, I began working with the BC Ministry of Agriculture. My first day there was an annual meeting at a hotel in North Vancouver and I met all the staff at one time — from Prince George, Creston, Kelowna, the Fraser Valley, and Vancouver. They were quite a crew, but I recognized right away that they were keenly interested in their role in regulating the dairy industry.

I was given a corner of an office in the basement of the Cassiar Street building (formerly a girls' reform school) and from there I went out somewhat timidly to fix the problems at the dairy plants. The first problem I encountered was the milk payment to farmers. For many years, plant lab staff measured the amount of butterfat with the very subjective Babcock Test. Infrared technology (known as IRMA) introduced a new way of measuring components in milk. The IRMA machines were huge, needing constant calibration and adjustments. Many people did not trust the new system and kept on using the Babcock. I would then sit down with plant personnel and compare our IRMA results to the Babcock. For a short time we used both systems. I would argue for the higher test on behalf of farmers; plant personnel would want the lower test to cover plant losses. Eventually, IRMA became accepted and payment to farmers became standardized, though for all the years I was with government there were anomalies that were not always explainable. Some of the problems we had were sample quality, frozen samples,

mislabeled samples, but it was a fairer way to pay dairy farmers.

They called me a Dairy Specialist, working mainly with the dairy plants. And at that time, there were a lot of them. I was assigned to the Lower Mainland and would visit all the dairies to obtain milk samples for analysis, review butterfat tests and do some plant inspection. I really enjoyed visiting the small dairies because of the stories I would hear. The White Lunch, a restaurant in downtown Vancouver, had a dairy plant called Acme Dairy, I think, on the third floor and accessed through the back alley and an industrial elevator. The entire dairy product was put up in stainless steel cans and sold in the restaurant on Pender Street, a great place to get a meal for just a few cents. There was also Jersey Dairy in Chilliwack, run by Hans Methorst, and Little Mountain Dairy in Abbotsford where the Waterhouse family had a busy dairy and coffee shop — a landmark in Abbotsford until it burned down. Shirley Farm Dairy in Delta belonged to the Malenstyn family and it was there that I met Bronco Horvath, who later became a dairy farm inspector. Many other dairies, including Palm Dairies and Jersey Farms, were all interesting and struggling to keep pace with technology and the demands of inspectors. I think they felt there were too many inspectors.

One very memorable plant I visited was in Ocean Falls. One day our office received a call from the public health nurse in Ocean Falls. She was convinced that the dairy there was making the people sick. Since the Dairy Branch of the Ministry of Agriculture had issued the licences for this plant to operate, I was asked to

Independent Milk Producers' Cooperative Association advertises milk from one of its affiliated dairies, 1930s.
GARDE GARDOM COLLECTION

Cyril Shelford, Minister of Agriculture, with Island Farms truck, early 1970s.
FRED MOCKFORD COLLECTION

Bottling milk at Island Farms, Victoria, 1940s.
BCARS I-02201

visit. The plant was really just a vat in the back of the grocery store where skim milk powder and unsalted butter were reconstituted into a product that resembled milk. The coliform counts were very high in the finished product and the conditions under which it was produced were very dirty. The store and plant operator was not usually sober and he was not prepared to clean up or make the necessary changes, so we closed the plant in support of the public health nurse. But that created a problem. There would be no milk to drink. So, upon returning to Vancouver, I went to see Neil Gray at Dairyland and they were able to ship milk into Ocean Falls on a regular basis.

Sometimes things happen at dairy plants that cause contamination to the product. This was a reason for visiting dairies. I remember in my zeal to find a problem and to solve it for Avalon Dairy, which is still a well respected dairy in Vancouver, Ev Crowley took great exception to what I had to say. He had no trouble going to the top and said he would phone the minister to fire me. I didn't lose my job and Ev and I eventually became friends.

Dairyworld Foods:
The Emergence of a National Dairy Cooperative

Dan Wong, 2000.

The 1990s saw big changes in the dairy industry in Canada. Dairyworld Foods was at the forefront of these changes.

David Coe succeeded Neil Gray as chief executive officer at the helm of the Fraser Valley Milk Producers' Association in 1986. For those with vision, there were unmistakable signals as early as the mid-1980s that heralded a long-term trend to deregulation. For dairy processors, controlled pricing and the relaxing of geographical boundaries within Canada were but two examples of the fundamental change which was about to take place. Dairyworld's vision of a national dairy cooperative quickly began to take shape in the early 1990s. Building on their fraternal roots, the dairy cooperatives in western Canada joined forces. In 1992, Fraser Valley Milk Producers' Cooperative Association merged with Edmonton's Northern Alberta Dairy Pool and with Red Deer's Central Alberta Dairy Pool to create a new western Canadian cooperative called Agrifoods International Cooperative Ltd.

Operating under the trade name Dairyworld Foods from its headquarters in Burnaby, Agrifoods became the third largest dairy supplier in Canada and the largest food manufacturer west of Ontario. The declared mission of Dairyworld Foods was to meet the needs of a new marketplace dominated by national customers. And in short order, it did — through the consolidation of operations in all provinces, investments in new plants and equipment, and the establishment of the Dairyland and Armstrong brands as market leaders throughout western Canada. Dairyworld's strategy of positioning itself in key markets across Canada made it ready for changes occurring on many fronts. By 1993, the long-standing tradition of not selling fluid milk across provincial borders had fallen apart in western Canada, and by 1994 discussions were underway to replace provincial milk pooling systems with a western Canadian pool.

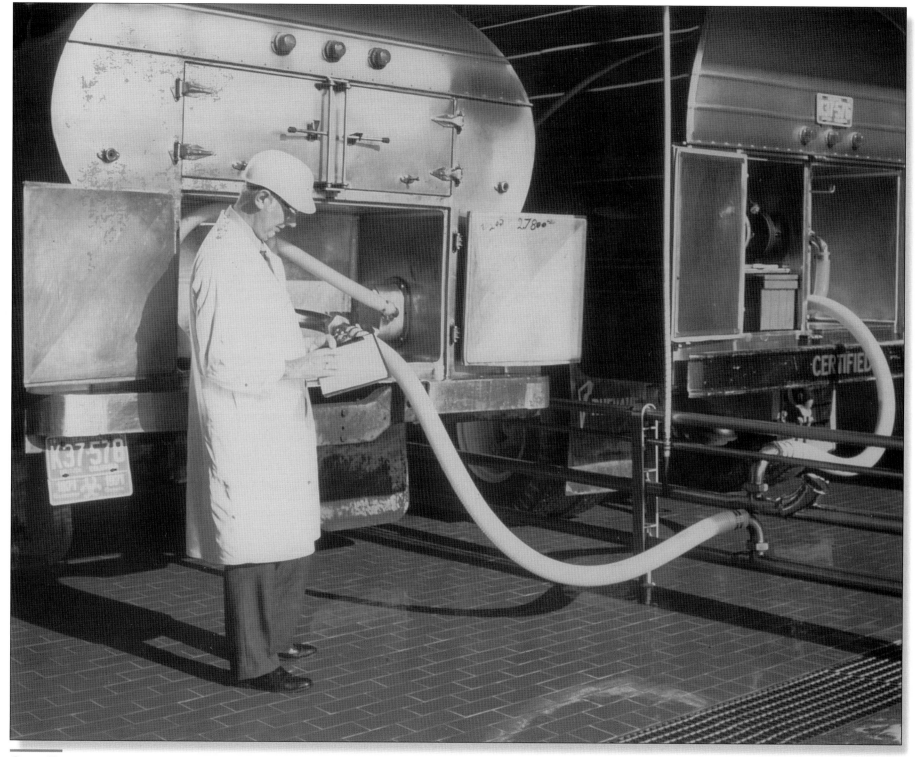

Gerry Vantongeren, laboratory supervisor, in milk receiving area of Lucerne's new Burnaby plant, early 1970s.

Enormous Change

When I became CEO, with the exception of our operation on Vancouver Island and a little reconstitution plant up in Kitimat, Dairyland was essentially a Lower Mainland dairy. I see myself as shepherding the organization through the enormous change to a truly national business, and finally, to a global business. There are two ways to compete in a market that is going global and is becoming more deregulated all the time: the first is to carve out a small niche — such as a geographic niche like Island Farms, or a specialty product niche like Avalon with its glass bottles. The second is to compete head on. If you're a mid-size dairy, which is what we were, you don't really have an option — you either have to shrink yourself down or grow to be able to take them on in their own court.

As a cooperative, one of our driving forces has been to protect our members. As this industry deregulates, the ultimate power is in the market so as long as we have a market presence we have a say in where our milk goes. When we sold our ice cream to Nestle, for example, we got a long term supply contract which is for the good of our members — it protects them. That's why we have to be what we are — to protect our long term interests.

David Coe, 2000.

David Coe

DAIRYWORLD COLLECTION

By entering the western pool, BC producers — including members of the former FVMPCA — began to share revenues and markets with their counterparts in Alberta, Saskatchewan, and Manitoba.

Dairyworld Foods knew that changes in the industry were not about to stop; even so, everyone in the industry was amazed by the extent to which change took place. By the late 1990s, almost half the national grocery market was in the hands of two chains. Among dairy suppliers, three organizations — including Dairyworld Foods — accounted for more than three-quarters of all the milk sold in Canada. Dairyworld Foods' expansion across Canada accelerated as it became apparent that British Columbia, like all provinces, was becoming more a component of the national market than a market unto itself. In addition, Canadian markets were being eyed by large multinational food companies. Unless companies like Dairyworld were able to operate on a similar scale, they would be frozen out of much of the national business driving the Canadian food industry. Through the steps it had taken, Dairyworld was prepared for the challenge.

In BC, venerable names like Silverwood Dairies, Palm Dairies, and Foremost Foods had disappeared with changes in the marketplace, and Dairyworld emerged as the leader in fluid milk and as the province's primary manufacturer of cheese. From its BC base, the company had also started to build a solid reputation exporting dairy products to markets outside Canada, primarily to Southeast Asia. At the same time, the Armstrong brand was growing in stature, and was judged to be the World's Best Cheddar at the World Championship Cheese Competitions in 1994 and 1998. Meanwhile, the consolidation of business operations required fewer, more sophisticated plants. Facilities were substantially upgraded in Burnaby, Abbotsford and Armstrong, while operations were discontinued in Vernon, Nelson, Prince George and Smithers. In 1997 Dairyworld Foods sold its ice cream manufacturing operations.

Dairyworld Food's national expansion between 1992 and 1999 included the acquisition of fluid processing operations in Ontario and Manitoba; a joint venture to market Yoplait yogurt and fresh dessert products nationally; a merger with Dairy Producers Cooperative Limited in Manitoba and Saskatchewan; and the creation of the Armstrong Cheese Company Ltd., a leading national cheese supplier. In 1998, Dairyworld teamed up with two small dairy cooperatives in Québec to form Nutrilait Inc., a supplier of refrigerated dairy products; it acquired an interest in Pascobel Inc., a Montréal-based dairy ingredients manufacturer; and, in 1999, it purchased Baxter Foods Limited, an established dairy processor in the Maritimes. Entering the twenty-first century, Dairyworld Foods was Canada's only truly national dairy supplier, with processing operations in eight provinces and product sales from coast to coast. ■

Official opening of new **NOCA** plant, 1967. The Honorable Pat Jordan salutes the farmers of the Okanagan.

Grand opening of Island Farms expansion after Silverwood purchase, 1983. On balcony, left to right: Ivor Fuller, Janet Baird, John Pendray, Reg Cottingham, Fred Mockford. Below balcony: Willard Michell (in wheelchair), Arthur Aylard, Gordon Ganong.

Conclusion

Winter scene, Fraser Valley, undated.
DAIRYWORLD COLLECTION

A few months ago Archie McNair offered to gather together a number of farmers and old-timers from the Richmond area to talk about their lives and their work in dairying. Archie now lives in White Rock, but he grew up in Richmond and farmed there, along the way witnessing astonishing changes in the municipality as urban growth edged out agriculture in this area so close to Vancouver. When we picked him up at his home on the morning of our gathering, he showed us that he had with him a list, drafted from memory, of Richmond producers — a document containing approximately 60 names. As we sat around listening, talking, and exchanging information that morning, those gathered — including Jim Holt Sr., Jim Holt Jr., Don May, Art Savage, Doug Savage, and Jake Savage — periodically punctuated the conversation with the addition of another name to Archie's list. By the time our two hours of our conversation were over, the list had swelled to 90 names and nobody was at all sure that it was complete. Today, only a handful of producers remain in Richmond.

But more milk is produced today than ever before. *Milk Stories* is about the shift in the dairy industry that Archie's list illustrates so well — the shift from many small farms at the turn of the century through radical changes in legislation, policy, and animal management in the fifties and beyond, to the shaping of a modern industry. *Milk Stories* has attempted to capture the flavour of life in the industry, and to document at least a part of the endless negotiations required to meet the challenges of living and working in different regions of British Columbia. It is just a beginning. ■

Milk Tokens

*We are indebted to Ron Greene of Victoria for loaning us the
images of milk tokens that grace the first page of each chapter.*

 Chapter 1 (page 1): Canadian Bank of Commerce Agricultural Fair medal. Awarded to A.C. Wells, of Edenbank Farm, at the Chilliwack fair, September 1911.

 Chapter 2 (page 23): Randolph Bruce medal awarded to the best young Ayrshire herd at the Armstrong Fair in 1929. Randolph Bruce was Lieutenant-Governor of BC.

 Chapter 3 (page 53): Registered Jersey Farms, Victoria, likely from the late 1930s.

 Chapter 4 (page 101): A.G. Carlson Dairy. This dairy operated in Revelstoke from 1902-1919.

 Chapter 6 (page 157): Jersey Dairy, Chilliwack.

 Chapter 7 (page 181): Frank Brothers Dairy, Terrace.

 Chapter 8 (page 241): Manor Farm, Saanichton.

 Chapter 9 (page 257): Balmoral Farm Dairy, Burnaby.

 Chapter 10 (page 279): A.G. Carter Dairy, Saanich.

 Conclusion (page 317): R.G. Millman

Acknowledgements

My work on this project has been indebted to a number of historians who have come before: Everett Crowley, of Avalon Dairy, who gathered notes on dairy history for many years; Morag Maclachlan for her foundational work on the history of the FVMPA and its context; Frank Bradley, who conducted interviews and gathered material for a history project in 1993 and 1994; Barbara Souter and Bill Howe for their work in gathering and preserving Holstein history at Westgen.

More recently, this project has gone forward through the stubbornness and vision of Peter Friesen, Neil Gray, and John Durham — who transformed a series of coincidences into the revival of a history project that had fallen dormant. Peter has been much missed in the latter stages of the project. John Durham, especially, deserves credit for the many hours of work he put in fundraising, making the project possible in a fundamental way. Many thanks also to the board members of the Dairy Industry Historical Society, all of whom contributed in important ways by drawing from their fields of expertise; to the many people who loaned materials and reminiscences to the project; to John Pendray and Des Thompson who worked above and beyond the call of duty interviewing and gathering material on Vancouver Island; to Sheila Durham, Tom Low, Neil Gray, Barbara Souter, Gordon Souter, Des Thompson, Sandra Ulmi, and Jacqueline Wood for reading the manuscript carefully; to Sandie Coplin, for helping with transcription of interviews; to the institutions who supported us and their staff who offered their help — especially Garde B. Gardom, Lieutenant-Governor of BC and his assistant, Suzanne Creighton; Priscilla Davis, Cowichan Valley Museum and Archives; Donna Redpath, North Peace Historical Society; Day Roberts, South Peace Historical Society Archives; Bev Kennedy, Kilby Farm Museum; Bryan Klassen, Langley Centennial Museum and National Exhibition Centre; Jim Armstrong, Dairyworld; Jim Byrne, BC Milk Marketing Board.

Thanks also to Anne Robertson for her archival assistance; to Marie and Dick Weeden, who loaned us material on Edenbank Farm that they plan to publish in the near future, (and who can be reached at Box 510, 108 Mile Ranch, BC, V0K 2Z0); to Rhiannon Gedak, for her patience and good humour as she photocopied pages by the thousand and transcribed tapes by the hour; to Al Tuchsherer and Alan MacInnes, researchers who came into the project late and brought with them a passion for asking questions about, and finding answers in, boxes of old papers ; to book designer Bill Glasgow, of Wm. Glasgow Design in Abbotsford, who created from my sketchy ideas and the contents of old boxes something to be proud of.

Thanks to Heather Watt, Mary Watt, and John Watt for being willing to step into child care at a moment's notice. And finally and always, to my husband, Greg Antle, for his unfailing support of my work, and to Tom and Georgie for their patience.

Notes

CHAPTER 1

1. Maclachlan 1998, 5.
2. Maclachlan 1998, 28-30.
3. Maclachlan 1998, 110.
4. Cullen 32.
5. Cullen 32.
6. Cullen 32.
7. Cullen 32.
8. Some historians believe that because Mr. Allard was a local storyteller who often told tall tales, the veracity of his accounts might be questioned.
9. Nelson 15.
10. Anderson 143.
11. Anderson 143.
12. Barman 43.
13. Laing n.p.
14. Innes 18.
15. Innes 18.
16. Mallandaine 20.
17. *Daily Alta California* (San Francisco) 23 April 1858. Cited in Sterne, page 2.
18. Molyneux 20.
19. Molyneux 20.
20. Barman 66.
21. Barman 68-69.
22. Barman 68-69.
23. Johnson 6.
24. Innes 215.
25. Innes 215.
26. Crowley 4.
27. Hibben 333.
28. Henry 35.
29. Hibben 346.
30. Hurford 2.
31. *A Century* 2.
32. Farrington and Woll 28.
33. Farrington and Woll 28.
34. *Commonwealth* (New Westminster) 15 September 1892. Cited in Innes, page 217.

CHAPTER 2

1. "A Brief History of Supply Management in Canada" 2.
2. Winter 111.
3. Freeman, preface.
4. Anderson 274.
5. *A Century* 4.
6. *A Century* 1.
7. Gilchrist n.p..
8. *A Century* 17.
9. Hare 6.
10. Hare 6.
11. Hare 6-7.
12. Souter 32.
13. Souter 32.
14. Souter 33.
15. BC Dairymen's Association Minutes, 1 March 1906.
16. BC Dairymen's Association Minutes, 10 January 1895.
17. BC Dairymen's Association Minutes, 2 April 1908.
18. Directors' Report to BC Dairymen's Association, 1908 or 1909.
19. *A Century* 7.
20. Ministry of Agriculture Annual Report, 1913. Pages 37-38.
21. Eagles 18.
22. University of British Columbia Undergraduate Information Guide, 1999-2000.

CHAPTER 3

1. *Our First Fifty* n.p.

2. Frank Bradley notes.
3. Maclachlan, FVMPA, 64.
4. *Our First Fifty* n.p.
5. Maclachlan, FVMPA, 56.
6. Maclachlan, FVMPA, 57.
7. Wamboldt 11.
8. Wamboldt 13.
9. Information about the history of Armstrong Creamery is from Blackburn, or from Dave Henley's autobiographical letter to Stan McHann, circa 1982, part of a collection of papers donated by Ernie Brown to the Dairy Industry Historical Society.
10. Information about Registered Jersey Farms is taken from a draft of Ron Greene's 1967 article by the same name.
11. Michell n.p.

CHAPTER 4

1. Clyne i.
2. Clyne 167.
3. Clyne i.
4. Clyne 165.
5. Clyne iii.
6. Maclachlan, FVMPA, 76.
7. Maclachlan, FVMPA, 79.
8. Maclachlan, FVMPA, 69.
9. Maclachlan, FVMPA, 81.
10. Maclachlan, FVMPA, 84.
11. Maclachlan, FVMPA, 95.
12. Maclachlan, FVMPA, 102.
13. Clyne iii.
14. Clyne 7.
15. Clyne xi.
16. Clyne xi-xii.

17. Clyne xii.
18. *Our First Fifty* n.p.
19. Clyne xi.
20. *Our First Fifty* n.p.

CHAPTER 6

1. British Columbia Milk Board, n.p.
2. History of the Canadian Dairy Commission, n.p.
3. British Columbia Milk Foundation, n.p.

CHAPTER 9

1. Snyder 4.
2. Snyder 11.
3. Snyder 222.
4. Snyder 11.
5. *Langley Advance*, 27 May 1948.
6. *Langley Advance*, 27 May 1948.
7. Snyder 226.

CHAPTER 10

1. Interview with David McMillan, 5 April 2000.
2. Jersey Farms pamphlet. Acton Kilby Papers, Kilby Farm Museum.
3. Jersey Farms pamphlet. Acton Kilby Papers, Kilby Farm Museum.
4. "Jersey Farms for Jersey Quality." Promotional brochure circa 1950. Acton Kilby Papers, Kilby Farm Museum.
5. Mutch 32.

Bibliography

Interviews

Conducted by Frank Bradley in 1993 and 1994

Arthur Aylard
A. Bates
David Blair
Doris Blair
Jean Crowley
Dorothy Davie
Bill Dinsmore
Ivor Fuller
Dick Graham
Neil Gray
Jeff Harbottle
Ruth Honeyman
Frank Norton
Bill Osborne
Maury Roberts
Marg Savage
Mac Sharpe
Geoff Thorpe

Conducted by John Pendray and Des Thompson in 1999 and 2000.

David Blair
Cyril Shelford
Dr. A. Kidd
Sig Peterson
Roy Wilkinson
Ann and Charlie Porter
Gordon Rendle
Bill Wilkinson

Conducted by Jane Watt in 1999 and 2000.

Bill Berry

Gordon (Chub) Berry
David Coe
Bob Cooper
Dorothy Davie
Fred Duck
Lloyd Duggan
Ivor Fuller
Garde B. Gardom,
Neil Gray
Ron Greene
Jim Holt Jr.
Jim Holt Sr.
Bob Irwin
Harry Irwin
Earl Jenstad
Grant Larkin
Dr. Joe Lomas
Don May
David McMillan
Keith Miller
Fred Mockford
Ruth Morrison
Frank Norton
John Pendray
Bill Ramsell
Archie McNair
Art Savage
Doug Savage
Jake Savage
Adrian Schrauwen
Gordon and Barbara Souter
Des Thompson
Allan Toop
Glenn Toop

Conducted by Gerry Adams, 2000.

Jim Calhoun

Bob Craig
Jim MacDonald
George Thom

Conducted by Al Tuchsherer, 2000.

Harry DeJong
Bill Howe
Gordon Park
Audrey and Barrie Peterson
Willoughby Rooke
Jim Waardenburg
Otto and Ruth Wuthrich

Primary Material from Private Collections

George Aylard
Darshen Bains
David Blair
BC Holstein Museum, formerly housed at Westgen in Langley.
Hilda Born
Frank Bradley
Ernie Brown
Fred Bryant
Jim Casanave
Jim Davidson
Andy Dolberg
Lloyd Duggan
Tom Erskine
Barb Ethier
Lorne Fisher
Garde Gardom
Walter Goerzen
Neil Gray
Ron Greene
Jim Holt Jr.

321

Bill Howe
Bob Irwin
Dr. A. Kidd
The Kooyman Family
Ray Kosiancic
Grant Larkin
Sandra May
Archie McNair
Bob Miller
Keith Miller
Fred Mockford
Harley Nicholson
Gordon Park
John Pendray
Jake Savage
Cyril Shelford
Wesley and Winnie Simpson
Walter Smith
Barbara and Gordon Souter
Harold and Kathy Steves
Allan Toop
Marie and Dick Weeden
Roy Wilkinson
Peter Wilson
Jack and Ellen Worrell
Dan Wong

Primary Material from Institutional Collections

BC Archives (BCARS)
BC Farm Machinery Museum
BC Milk Producers' Association
City of Vancouver Archives
Cowichan Valley Museum and Archives
Dairyworld Foods
Island Farms
Acton Kilby Papers, Kilby Farm Museum

Langley Centennial Museum and National Exhibition Centre
Naniamo Community Archives
North Peace Historical Society
Prince George Public Library
Quesnel and District Museum and Archives.
Richmond Museum and Archives
Saanichton Quarantine Station
South Peach Historical Society Archives
University of British Columbia – Special Collections
Vancouver Public Library
Westgen

Journals and Newspapers

Agricultural Journal (an early publication of the BC Department of Agriculture)

The British Columbia Magazine

Butter-Fat (in-house magazine of the Fraser Valley Milk Producers' Association)

Canadian Guernsey Breeders Journal

The Cariboo Observer

The Cream Collector (in-house magazine of the Shuswap Okanagan Dairy Industry Cooperative Association)

Crofton Gazette

Delta Times

The Holstein-Friesian Journal

The Industry formerly called *The Dairy Industry*

Island Farm News (early in-house magazine of Island Farms)

Langley Advance

Man-To-Man

NOCA Pictorial

Pure-Pak News Pictorial

The Vancouver Province

Other Materials

"A Brief History of Supply Management in Canada." *Industry Insider.* 1999.

A Century of Achievement. Victoria: Ministry of Agriculture, Fisheries, and Food, 1996.

Agassiz Research Station, 1886-1986. By Jack A. Freeman. Historical Series Number 33. Ottawa: Research Branch, Agriculture Canada, 1986.

Anderson, James Robert. "Notes and Comments on Early Days and Events in British Columbia, Washington and Oregon: Including an account of sundry happenings in San Francisco, being the memoirs of James Robert Anderson." Typescript, BCARS Add mss 1912 [c1914-1925].

Barman, Jean. *The West Beyond the West: A History of British Columbia.* Toronto: University of Toronto Press, 1991.

Berry, J.W. "Annual Statement, 1920." *Our First Fifty: Fraser Valley Milk Producers' Association, 1917-1967.* Burnaby: FVMPA, 1967.

Blackburn, Mary. "Armstrong Cheese Cooperative Association: 1939-1961," *Okanagan Historical Society Forty-Seventh Report.* 52-56.

British Columbia Dairymen's Association. Meeting Minutes. 30 January 1895. BCARS GR510.

British Columbia Dairymen's Association. Meeting Minutes 1 March 1906. BCARS GR510.

British Columbia Dairymen's Association. Minutes of Annual General Meeting. 2 April 1908. BCARS GR510.

British Columbia Dairymen's Association. Director's Report. [1908 or 1909]. BCARS GR510.

British Columbia Milk Board Consolidated Order. 1 August 1999.

British Columbia Milk Foundation. Brochure, 1970. Sandra May Collection.

Canadian Dairy Commission. Information retrieved from website 16 February 2000.

Carr, E.C. A Decision of the Milk Board, 16 April, 1947. George Aylard Collection.

Church, H.E. *An Emigrant in the Canadian Northwest.* London: Methuen, 1929.

Clyne, J. V. *Royal Commission on Milk.* End of citation.

Cross, S.W. "Report of the Dairy Committee to the 39th Annual Convention of the BC Federation of Agriculture," 1972. BC Milk Producers Association Collection.

Crowley, Everett. "The Beginning of a Dairy Industry in BC." In *The Story of Avalon Dairy Ltd.* By Jean Crowley. Vancouver: Avalon, 1996.

Crowley, Jean. *The Story of Avalon Dairy Ltd.* Vancouver: Avalon Dairy, 1996.

Cullen, Mary K. "The History of Fort Langley 1827-96." *Canadian Historical Sites Occasional Papers in Archaeology and History.* 20 (1979): 5-22.

"Dairying in the Bulkley Valley." Typescript. [1931] BCARS GR509. Department of Agriculture, Diary Branch papers, 1912-1947.

Department of Agriculture. *Annual Report.* Victoria: 1913.

Dominion Experimental Farm, Agassiz BC: Interim Report of the Superintendent, W.H. Hicks, For the Year 1921. Ottawa: Dominion of Canada, Department of Agriculture, 1922.

Eagles, Blyth. "Agricultural Extension at UBC Past and Present." Eagles Fonds, UBC Special Collections, 1961.

Evans, Robert. Cited in an article by E. Blanche Norcross in the *Cowichan News Leader,* 17 October 1956.

Farrington, E.H. and F.W. Woll. *Testing Milk and Its Products: A Manual for Dairy Students, Creamery and Cheese Factory Operators, Food Chemists, and Dairy Farmers.* 20th Edition. Madison WI: Mendota Book Company, 1911. BC Farm Machinery Museum Collection.

Friesen, Peter. "President's Report to the Fraser Valley Milk Producers' Cooperative Association," 1986. Dairyworld Collection.

Fifty Years of Progress on Dominion Experimental Farms: 1886-1936. Ottawa: Dominion Department of Agriculture, 1939.

Freeman, Jack A. *Agassiz Research Station: 1886-1986.* Historical Series Number 33. Ottawa: Research Branch, Agriculture Canada, 1986.

Gardom, Basil. "Compulsion in Primary Industry is Loss of Free Citizenship." Radio Address over CBR, 26 November 1937. Reprinted in *The Industry,* 1937. Garde Gardom Collection.

John E. Gibbard. "Early History of the Fraser Valley, 1808-1885." University of British Columbia, MA thesis, 1937.

Gilchrist, M.M. "A History of the British Columbia Department of Agriculture." Typescript, 1966. Roy Wilkinson Collection.

Grauer, Jake. "An Interview with D. Cleland," 1973. Richmond Museum and Archives.

Hare, H.R. "Memorandum of the Economic Factors of Butter Fat Production in BC." Typescript [1929]. BCARS GR509. Department of Agriculture, Dairy Branch papers, 1912-1947.

Harris, Cole, and Elizabeth Phillips, eds. *Letters from Windermere, 1912-1914.* Vancouver: University of British Columbia Press, 1984. Reprinted 1991.

Harris, J.C. "Memories of a Cowichan Landmark." *The Cowichan Leader,* 7 March 1946.

Hazelmere Women's Institute, Declaration of Association under Agricultural Associations Act, 1911, 2 July 1912. Langley Centennial Museum and National Exhibition Centre.

Henry, Tom. *Small City in a Big Valley: The Story of Duncan.* Madeira Park: Harbour, 1999.

History of the Canadian Dairy Commission." Canadian Dairy Commission Website. *Http://cdc.ca/history.* Retrieved 16 February 2000.

Hulbert, John. "Report of the Dairy Industry to the 36th Annual Convention of the BC Federation of Agriculture, " 1969. BC Milk Producers Association Collection.

Hurford, R.U. "Dairying in the Comox District: And Early History of the Dairying Industry in the Comox Valley is Closely Associated with the Activities of Mr. Alex Urquhart, who is now Retired from Business." Typescript, undated. BCARS GR509.

Ingledew, N.H. Letter to Island Farms members, 19 January, 1949.

H.A. Innes, Ed. *The Dairy Industry in Canada.* Toronto: Ryerson, 1937.

Johnson, Edward Philip. "The Early Years of Ashcroft Manor." *BC Studies* 5 (1970): 3-23.

Kendall, Wally. Quoted in 'The Milk Always Gets Through." By Barry Mather. *The Story of Dairyland.* Vancouver: Dairyland, 1950: 11.

Laing, F.W. "Early Agriculture in BC." Department of Agriculture, Dairy Branch Papers 1912-1947. Typescript, 1 August 1925, BCARS GR509.

Loney, Austin. Quoted in "The Milk Always Gets Through." By Barry Mather. *The Story of Dairyland.* Vancouver: Dairyland, 1950: 10.

Luckhart, Grace. "He Can't Win." *The Story of Dairyland.* Vancouver: Dairyland, 1950: 21-23.

Maclachlan, Morag. "The Fraser Valley Milk Producers' Cooperative Association: Successful Cooperative." MA Thesis, UBC, 1972.

Maclachlan, Morag. Ed. *The Fort Langley Journals, 1827-30.* Vancouver: University of British Columbia Press, 1998.

Maclachlan, Morag. "The Success of the Fraser Valley Milk Producers' Association." *BC Studies* 24 (Winter 1974-75): 52-64.

Mallandaine, Edward. *First Victoria Directory.* Victoria: Mallandaine, 1860.

Maloney, Mamie. "Beyond the City Limits." *The Story of Dairyland.* Vancouver: Dairyland, 1950: 25-26.

Michell, W.W. "President's Report." Typescript 1964. George Aylard Collection.

Mutch, Jean. "Not Just a Beverage." *The Story of Dairyland.* Vancouver: FVMPA, 1950: 31-33.

Grant MacEwan. Quoted in *BC Holsteins: 1886-1983.* By Barbara Souter. BC Branch, Holstein-Friesian Association of Canada, 1983.

McIntyre, James. *Oh! Queen of Cheese: Selections from James McIntyre, the Cheese Poet.* With additional cheese pieces by Roy A. Abrahamson. Toronto: The Cherry Tree Press, 1979.

McKinnon, Donald. Taped reminiscences, 1993. Island Farms Collection.

Macken W.L. "Address to FVMPA Annual Picnic," 8 July, 1939. In *Our First Fifty: Fraser Valley Milk Producers' Association, 1917-1967.* Burnaby: FVMPA, 1967.

Memories Never Lost: Stories of the Pioneer Women of the Cowichan Valley. Duncan: Pioneer Researchers, 1986.

Minutes of BC Federation of Agriculture Directors Meeting, 27 January 1953. BC Milk Producers Association Collection.

Minutes of Directors Meeting, Registered Jersey Dairies, 13 October 1939. Fred Mockford Collection.

Molyneux, Geoffrey. *British Columbia: An Illustrated History.* Vancouver: Polestar, 1992.

Nelson, Denys. *Fort Langley 1827-1927: A Century of Settlement.* Vancouver: Art, Historical, and Scientific Association of Vancouver, 1947.

Old-Time Agriculture in the Ads. By Robert F. Karolevitz. Aberdeen, SD: North Plains Press, 1970.

Our First Fifty: Fraser Valley Milk Producers' Association 1917-1967. Burnaby, BC: FVMPA, 1967.

Otter District Women's Institute, Declaration of Association under Societies Act, 16 August 1921. Langley Centennial Museum and National Exhibition Centre.

Park, W.J. "Memoirs." Typescript, 1956. Gordon Park collection

Paull, Elmore T., "Reminiscing About Holsteins" in *B.C. Holsteins 1886-1983.* Ed. Barbara Souter. BC Branch, Holstein-Friesian Association of Canada, 1983: 37-38,59-64.

Peterson, Barrie. "Report to the BC Federation of Agriculture Annual General Meeting," 1976. BC Milk Producers Association Collection.

Rive, H. "We Are Not Milking Cows for Honor and Glory–What for then?" Typescript [1939 or 1940] BCARS GR509. Department of Agriculture Dairy Branch papers, 1912-1947.

Sears Roebuck and Company Catalogue, 1908. Reprint. Chicago: Follett Publishing Company, 1969.

Shelford, Cyril. "An Overview of the British Columbia Dairy Industry: Prepared for the BC Milk Board," 1988. Cyril Shelford Collection.

Snyder, Roy G. *Fifty Years of Artificial Insemination in Canada, 1934-1984.* Guelph, Ontario: Canadian Association of Animal Breeders, 1984.

Souter, Barbara. *BC Holsteins, 1886-1983.* BC Branch, Holstein-Friesian Association of Canada, 1983.

Toop, Fred. Interviewed in 1963 by Imbert Orchard. In *Floodland and Forest: Memories of the Chilliwack Valley.* Victoria: Provincial Archives of BC, 1983.

Walker, H.E. In *A Century of Achievement.* Victoria: Ministry of Agriculture, Fisheries, and Food, 1996.

Wamboldt, Beryl. *The History of SODICA.* Vernon: Vernon News, 1965.

Wejr, Stan. "Memories of Trinity Creek Area in the 1920s." *Okanagan History* 50 (1986): 78-79.

Wells, Oliver, 1967. "Edenbank: The Story of a Family Farm." Typescript, 1967. Revised by Marie and Dick Weeden, January 1999. Marie and Dick Weeden Collection.

Winter, George R. "Agricultural Development in the Lower Fraser Valley." *Lower Fraser Valley: Evolution of a Cultural Landscape.* Ed. Alfred H. Siemens. Vancouver: Tantalus Research, 1968: 101-115.

Yale, James Murray. Letter to George Simpson, 1839. In Mary K. Cullen, "The History of Fort Langley 1827-96." *Canadian Historical Sites: Occasional Papers in Archaeology and History.* 20 (1979): 33.

York, Mrs. T. "Memories." From Frank Bradley notes. Dairy Industry Historical Society of BC Collection.

Sponsors

First SODICA Green Pasture Tour, circa 1954.

IN RECOGNITION

These dairy industry honorees are remembered by families and friends.

Honorees   *Remembered by*

Walter Adams, (1906–1988), Can Hauler from 1925 to 1969, Matsqui Prairie. . . . . Mabel Adams

Lyle Atkinson, (1899–1987), FVMPA Manager, 1961-1966. . . . . A friend

Arthur W. Aylard, (1904–1995), Producer, Saanich. . . . . George R. Aylard

Albert Bartels, (1924–2000), Producer, Sumas Prairie. . . . . George Bartels

John W. Berry, (1872–1943), Producer, Langley. . . . . W.J. & G.H. Berry, Belmont Farms Ltd.

Harry S. Berry, (1901–1980), Producer, Langley. . . . . W.J. & G.H. Berry, Belmont Farms Ltd.

William (Bill) Billeter, (1895–1984), Producer, Bulkley Valley. . . . . Jim Davidson, Canyon Creek Farm Ltd.

Irvine Blair, 28 years driving for Rooke's Cartage, Langley. . . . . Willoughby C. Rooke

Jacob W. Born & Hilda Born, Producers, Mount Lehman. . . . . John and Ellie Born, Village Ranch Ltd.

William Born (1897–1969), & Katerina Born (1907–1986), Producers, Matsqui. . . . . Jacob and Hilda Born

Frederick Paul Bryant, (1952–2000), Producer, Rosedale. . . . . Fred H.Bryant

Alan Burwash, (1904–1996) Manager of Jersey Farms, 1942-1968. . . . . Frank Bradley

Jim Davidson, Producer, Bulkley Valley. . . . . Rick Vandenberg, Vandenberg Dairy

G.K. Devitt, (1882–1971), Producer, Barnston Island. . . . . Mrs. Dagny Yurkin, Fraser Isle Farms Ltd.

Honorees   *Remembered by*

Albert Doney, (1897–1967), Producer, Saanichton. . . . . George and Christine Doney
Andy Driesen, (1940–1996), Producer, Abbotsford. . . . . Jonathan Driesen, Vulnaho Dairy
James Erskine, (1884–1974), Producer, Richmond - Delta. . . . . Thomas Erskine and E. Erskine
Sidney Fox, (1894–1968), Producer, Saanichton. . . . . Jim Fox, Silver Rill Farm Ltd.
Stan Fox, (1924–1995), Producer, Saanichton. . . . . Gordon J. and Heather Fox, Silvervale Farm
Peter Friesen, (1925-1999), Producer & Leader, Abbotsford. Agriculture Committee, Abbotsford Chamber of Commerce
James T. (1884–1980) & Janet Godfrey (1888–1969), Producers, Saanichton. . . . . Gordon Godfrey
Alexander Graham, (1909–1999), Producer, Sumas Prairie. . . . . David Graham, Graham Dairy Farms Ltd.
Joe & Joanne Groenendijk, Producers, Chemainus. . . . . Chris Groenendijk, Greendike Farm Ltd.
Thomas P. Harbottle, (1879–1972), Distributor, North Vancouver. . . . . Jeffery P. Harbottle
Gerald R. Hennel, Can Hauler from 1951 to 1965, Langley. . . . . Gerald R. Hennel
Lambert & Shirley Hylkema, Producers from 1956 to 1993, East Chilliwack. . . . . Francis J. Hylkema, Rypster Farm
John Johnson, (1882–1963) & Olga Johnson (1889–1967), Producers, Creston Valley. . . . . Earl and Auriol Jenstad
Jager Singh Judge, (1904–1976), Producer, Victoria. . . . . Kalla Judge
Wiebe Keulen, (1927–1995), Producer, Delta. . . . . Art and Wilma Keulen, Neveridle Farms
John Mackay Kirkness, (1905–1990), Producer, Sardis. . . . . Marinus Schryver, M.C.M. Holsteins Ltd.
John Klassen, (1901–1960), & Maria Klassen, Producers, Matsqui. . . . . Jacob and Hilda Born
Ray Kosiancic, Producer, Crescent Valley. . . . . Ron and Sandy Ulmi
Jim A. MacDonald, (1882–1954), Can Hauler, Mount Lehman. . . . . James R. MacDonald
John Wesley McGillivray, (1874–1932), Producer, Chilliwack. . . . . Edith Toop
Barry D. McKnight, Chilliwack. . . . . KPMG, (Hank Kroeker)
Arnold Mammel, (1923–1999), Producer, Chilliwack. . . . . Gerry and Jan Mammel
Jacob G. Martens, (1908–1981), Producer, Chilliwack. . . . . George Martens, Delmar Dairy
Alex Mercer, (1878-1961), FVMPA Manager, 1933-1961. . . . . A friend
Robert Mercer, (1892–1988), Producer, Saanich. . . . . Phyllis G. Roberts
Willard W. Michell, (1899–1986), Producer, Saanichton. . . . . Maurice and Veda Michell
Ruth Morrison, long-time executive secretary, FVMPA. . . . . A friend
James (Jim) Munro Sr., (1854–1920), Producer, Rosedale. . . . . Mildred A. (Munro) Picket
George J. Okulitch, (1909–1980), FVMPA Manager, 1960s. . . . . A friend
G.W. (Bill) Ormel, (1933–2000), Producer, Fraser Valley. . . . . Chris Ormel, Orbo Estates (1973) Ltd.
William James Park, (1880-1964), Producer, Pitt Meadows. . . . . Gordon Park
John H. Poelman, (1958–1996), Producer, Cobble Hill. . . . . Poelman Family, Sunny Vale Farm Ltd.
Charles Pendray, (1893–1966), Producer, Saanich. . . . . John Pendray and Family, Pendray Farms Ltd.
Thomas Pendray, (1902–1981), Producer, Saanich. . . . . Robert T. Pendray
Ralph Rendle, (1892–1989), Producer, Victoria. . . . . Gordon Rendle, Stanhope Dairy Farm Ltd.
David Rennie, (1926–1950), Producer, Matsqui. . . . . Doug and Alice Rennie, Marcath Farms
Thomas J. Robertson, (1892–1963), FVMPA Director, Delta. T.W. (Bill) Robertson
Willoughby C. Rooke, Independent Milk Hauler, Langley. . . . . Richard and Marion E. Hocking
Paolo Scardillo, (1921–1992), Cheese Maker, Vancouver. . . . . Rocky and Kathy Scardillo
David Skea, (1903-1960), Producer, Langley. . . . . Steve Wurst
Mr. G. H. Singh, (1891–1956), and Mrs. Singh, Producers, Chilliwack. . . . . Mr. & Mrs. G.J. Singh Bains, Bains Dairy Farm Ltd.
William Henry Street, (1926–1983), Producer, Victoria. . . . . Ralph Street, Peninsula Bulldozing Ltd.
Gordon A. Swan, (1897–1983), Producer, Saanich. . . . . Len & Marg Swan
Forrest Edward Telford, (1911–1992), Producer, Aldergrove. Jim Telford, Telford Farms
John Tenbrinke, (1925–1991), Producer, East Chilliwack. . . . . Grace-Mar Farms
Albert Edwin Toop (1890–1958), & Annie Toop (1893–1992), Producers, Chilliwack. . . . . Muriel Toop
Alvin Toop, (1915–1967), Producer, Sardis. . . . . Sandra Andreasen
Gordon H. & Annie Toop, (1905–1987), Producers, Chilliwack. . . . . Robert G. Toop
Ellwood D. Toop, (1903–1979), Producer, Chilliwack. . . . . Richard D. Gill and Patricia Gill
Norman E. Tupper, (1922–1992), FVMPA. . . . . A friend

Honorees   *Remembered by*

Nelly Van den Dungen, (1926–1998), Producer, Surrey. . . . . William Van den Dungen
Harke Vandermeulen, Bulkley Valley. . . . . Rick Vandenberg, Vandenberg Dairy
Adrie and Quirinus van Dongen, Producers, Delta. . . . . John van Dongen
Waardenburg Brothers Farm, owned and operated by
Wayne, James, and Wesley Waardenburg and Paul Van Dokkumburg. . . . . Jim Waardenburg
Jan Willem Wikkerink, (1907-1993), Producer, Cobble Hill. . . . . Rudolf & Sadie Wikkerink,
Father Thomas Wilkinson, (1873–1956), Producer, Cobble Hill. . . . . William & Beatrice Wilkinson
William H. (Bill) Wilkinson, (1917–2000), Producer, Cobble Hill. . . . . Shincliffe Enterprises Ltd.
Francis Vanstane Worrell, (1856–1927), Producer, Milner. . . . . Ellen Worrell
George Worrell, (1889–1960), Producer, same farm in Milner. . . . . Ellen Worrell
John Francis Worrell, Producer, same farm in Milner. . . . . Ellen Worrell
Thomas R. Youell, (1911–1988), Producer, Saanich. . . . . Youell Bros., Youelldene Farm
Andy Yusko, Producer, Pitt Meadows, Local Secretary, FVMPA. . . . . Mike Yusko

---

Milk Stories Contributors

PLATINUM PLUS SPONSORS

Province of British Columbia, BC2000 Millennium Arts & Heritage Fund
British Columbia Dairy Foundation
British Columbia Milk Marketing Board
Dairyworld Foods

PLATINUM SPONSORS

Avalon Dairy Ltd.
Borden Ladner Gervais LLP,
B.C. Milk Producers Association
Clearbrook Grain & Milling Co. Ltd.
Ritchie-Smith Feeds, Inc.
Tetra Pak Canada Inc.
Top Shelf Feeds Inc.
Westgen

GOLD SPONSORS

Agropacific Industries Ltd.
Curwood Packaging (Canada) Ltd.
KPMG LLP
William M. Mercer Limited

PATRONS OF THE SOCIETY

Abbotsford Veterinary Clinic Ltd.
Ball Packaging Products Canada, Inc.
Birchwood Dairy's Inc.
Burgess Agri Supplies Ltd.
Continental Colloids Canada Inc.
Co-operators Insurance
Dairy Farmers of Canada
Farm Credit Corporation
Greenbelt Veterinary Services
Island Farms Dairies Co-operative Association
Loewen Welding & Mfg. Ltd.
Lucerne Foods, (A Division of Canada Safeway Ltd.)
Macaulay McColl, Barristers
Mainland Dairymen's Association
Murphy & Wakefield Ltd.
Natrel Inc. (SealTest)
Otter Co-op
Pendray Farms Ltd.
Portola Packaging Canada Ltd.
Saputo Group Inc.
Amcor Twinpak - North America Inc.
Ty-Crop Manufacturing Ltd.
Vedder Transport Ltd.

FRIENDS OF THE SOCIETY

Abbotsford Chamber of Commerce
Alder Vista Farm Ltd.
Associated Labels
Avenue Machinery Corp.
Baker Newby , Barristers & Solicitors
Bank of Montreal Group of Companies
Barton Insurance Brokers Ltd.
Bobcat Country Inc.
Delaval J&D Farmers DAiry Service Ltd.
Ecolab Ltd., Food & Beverage Division

Enhance Packaging Technologies Inc.
Grober Inc.
HSBC Bank Canada
Landmark Dairy Ltd.
National Dairy Council of Canada
Rentway Ltd., (Penske Truck Leasing)
Royal Bank of Canada
Telford Farms Ltd.
Valley Equipment Ltd.
Vopak, (Van Waters & Rogers Ltd.)

DONORS

Arc Appraisals
Canada Colours and Chemicals Ltd.
Chilliwack Chamber of Commerce
Creston Valley Farms
Samuel L. Francis
Robert Hrabinsky
Keith Miller
Pacific Dairy G.S. Centre Ltd.
Barrie and Audrey Peterson
Purity Packaging Ltd.

Ridgebeam Farms Ltd.
Richard Savoury, Glopak
Shuswap Veterinary Clinic
Walter and Maria Smith
Sunny Vale Farm Ltd.
United Grain Growers Ltd.
UAS Industries
Terral Farms Incorporated
Dr. Brian Upper, Elkview Veterinary Clinic
Valley Farm Drainage

Borden Ladner Gervais LLP is proud to support the Dairy Industry Historical Society

BRITISH COLUMBIA

2000

Marking The Millennium

The Province of British Columbia is pleased to support the Dairy Industrial Historical Society of B.C. through our BC2000 Arts and Heritage program. This historical and visual account will draw on the wealth of knowledge and experience in rural B.C.

The dairy industry is a significant contributor to BC's economy, creating jobs, secondary industries and providing top-quality products. This celebration of the millennium captures the stories of our pioneers, and preserves and interprets the history of one of the largest sectors of our community for today and the future. This is a true community project.

I applaud the society for its excellent work and for publishing this wonderful documentary.

Andrew Petter
Minister Responsible for
British Columbia 2000

B.C. owned and operated.
Growing with the farm community since 1968.

ritchie-smith feeds, inc.

33777 Enterprise Avenue
Abbotsford, B.C. V2S 7T9
Phone: 604-859-7128 1-800-242-8011 Fax: 604-859-7011

Through world-renowned nutrition education, school milk programs, and innovative advertising, BC Dairy Foundation plays a significant role in promoting milk and dairy products to the people of British Columbia.

BC Dairy Foundation is proud to have been a part of the BC dairy industry for over a quarter of a century.

BC Dairy Foundation Presidents of the Board of Directors

Gordon Park 1974	Peter Friesen 1980-1981	Stan Cross 1988-1989	Louis Yonkman 1996-1997
Herb Argue 1975	Harley Jensen 1982-1983	Harley Jensen 1990-1991	Dave Blackwell 1998-1999
John Pendray 1976-1977	Jim Davidson 1984-1985	John Van Dongen 1992-1993	Randy Kitzel 2000-2001
Bill Hadwell 1978-1979	Gordon Ganong 1986-1987	Barry Trojanoski 1994-1995	

BC DAIRY FOUNDATION

Tetra Pak

More than the package.

Leaders in packaging and processing for the liquid food industry for over 49 years.

The origin of the Tetra Pak Processing Division goes back to a man called Gustaf de Laval who invented the first "continuously operating cream separator." The year was 1877.

Based on this invention AB Separator was founded 1883. The company later on changed name to Alfa-Laval and in 1993 Tetra Pak acquired the company. The Alfa Laval food and dairy processing division was merged in to Tetra Pak's organization which made Tetra Pak the leading Dairy Processing and Packaging company around the world.

Tetra Pak Canada Inc.

10 Allstate Parkway
Markham, Ontario, L3R 5P8
Telephone: (905) 305-9777
Telefax: (905) 305-6900

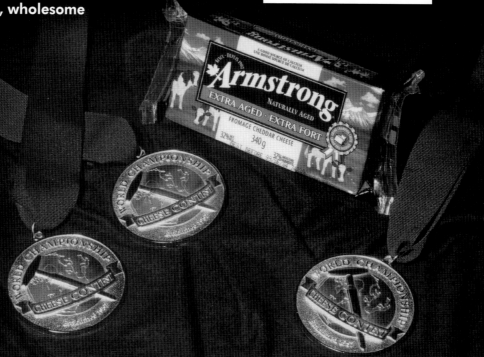

We've grown because of you.

While Westgen is proud to have grown and developed with BC Agriculture, we look forward to a continued leadership role of meeting the A.I. needs of BC's Beef and Dairy Producers.

WESTGEN

Box 40
Milner, BC V0X 1T0
Ph.: 604-530-1141 Fax: 604-534-3036
www.westgen.com

Avalon DAIRY LTD.

HOME OF ORGANIC CERTIFIED Milk B.C.'S FIRST

SINCE 1906

B.C. Milk Producers ASSOCIATION

2509 Vancouver Street, Victoria, B.C. V8T 4A6
phone: 250-383-7171 fax: 250-383-5031
e-mail: dolberg@telus.net

LOOKING BACK... The early history of the BC Milk Producers Association was to a large extent a story of the initial years of the BC Federation of Agriculture. The founding president of the Federation (incorporated as the BC Chamber of Agriculture in 1936) was Fraser Valley Milk Producers Association visionary, dairyman, and former agriculture minister, the Honourable E.D. Barrow. While the FVMPA was the only founding dairy organization member, the federation's membership grew over the years to include all regional dairy associations.

The BCFA served as a forum for the dairy groups from across the province to develop consensus on a wide range of dairy policy issues. Under the chairmanship of Archie Stevenson, the BCFA Dairy Committee became formalized as a standing committee in the early 1960's, and was subsequently chaired by the following dairymen:

Len Bawtree	1966
Gordon W. Park	1967
John Hulbert	1968-69
Len Bawtree	1970
Tom K. Berry	1971
Stan W. Cross	1972-73
Barrie Peterson	1974-75
George Aylard	1976-78
Harke Van der Meulen	1979
Barrie Peterson	1980
Harke Van der Meulen	1981-82
Melle Pool	1983-84
Edgar Smith	1985-86

The successor to the Dairy Committee, the BC Federation of Dairymen's Associations, was incorporated in 1987 to serve as a forum for regional milk producer associations to develop and pursue policies and programs beneficial to the BC dairy industry. In 1995, the BCFDA changed its name to the BC Milk Producers Association. The organization has been chaired over the years by the following producers:

August Bremer	1987-88
John van Dongen	1989-90
Johanna Mellor	1991-92
Ben Brandsema	1993
Anne Lang-Harris	1994-95
Ben Cuthbert	1995-96
Bill Park	1996
Tom Nash	1997-99
Wally Smith	1999-2000

LOOKING FORWARD... Reflecting on the past accomplishments documented in *Milk Stories* establishes the importance of continuing to work together to ensure a bright future for the dairy industry. As such, the BCMP maintains a positive working relationship with Mainland Dairymen's Association, the BC Milk Marketing Board and other dairy industry partners. Through the successor of the BCFA, the BC Agriculture Council, the association works collectively with all other BC commodity groups to pursue common objectives. At the national level, the BCMP participates in the policy development process through both Dairy Farmers of Canada and the Canadian Federation of Agriculture.

These collective efforts have paid off in the past and will indeed provide for prosperity into the future for dairy producers from across British Columbia.

Feed mill construction on Enterprise in 1965.

Existing feed mill operation, 2000

CROWN FEEDS BRAND

Clearbrook Grain & Milling Co. Ltd.

2425 TOWNLINE ROAD, P.O. BOX 2400 ABBOTSFORD, B.C. V2T 4X3 • OFFICE: (604) 850-1108 • FAX: (604) 850-3854

Servicing local farmers since 1953

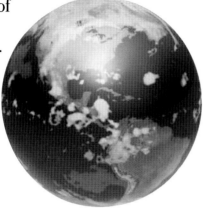

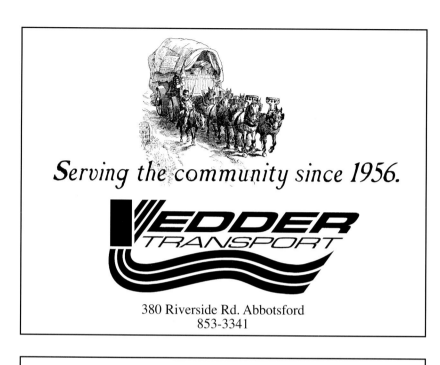
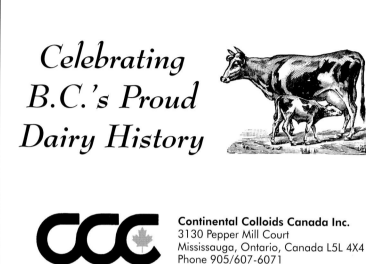

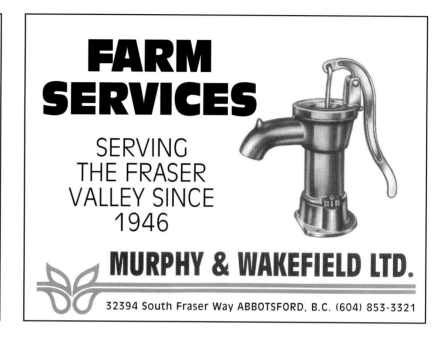

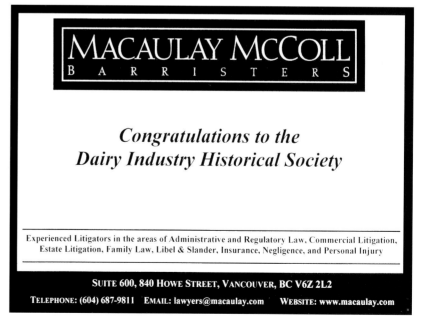

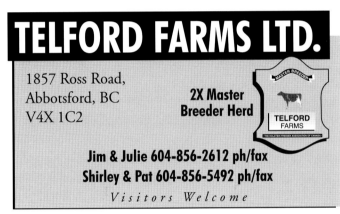
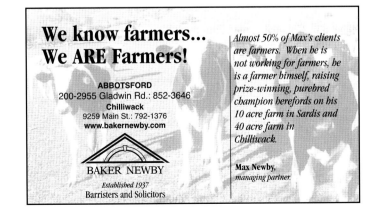

Index